Wave Mechanics and Wave Loads on Marine Structures

Wave Mechanics and Wave Loads on Marine Structures

Editor

Alfred Eckstein

scitus
academics

Wave Mechanics and Wave Loads on Marine Structures
Edited by **Alfred Eckstein**

ISBN: 978-1-68117-575-1
Library of Congress Control Number: 2016943010

SCITUS Academics LLC,
www.scitusacademics.com
Box No. 4766, 616 Corporate Way,
Suite 2, Valley Cottage,
NY 10989

Printed in United States of America on Acid Free Paper

Preface

Marine structures can be of different types of engineering conveniences that are designed and constructed for marine resource utilization and constant improvement in the ocean. Analysis, design and construction of offshore structures could be one of the most demanding tasks faced by the engineering industry. In addition to the common situations and conditions encountered by land-based structures, offshore structures have the added complication of being placed in an ocean environment in which hydrodynamic interactions and powerful reactions begin to be important considerations in their design. Weight control is critical to an offshore structure as the construction of it is governed by a strict criterion of principles.

This book is intended to present insights of some of the key factors in the analysis and design of offshore structures that an engineer uninitiated in offshore engineering should consider. The topics covered range from the theories of water waves, structure-fluid interface in waves to forecasting extreme response values from spectral modeling. A simple understanding of a variety of key topic areas is crucial for the design of offshore structures by an engineer. It presents a new outlook on measuring wave forces on ocean structures, combining deterministic and probabilistic methodologies with theory of waves and linking field and experimental measurement methods.

Table of Contents

1

CHAPTER

The Wave Properties of Zuara Coast and Their Effects on the Marine Navigation

Jamal Algraeo[1*], Samir Bouaziz[2]

[1]*Geography Department, College of Arts, Azzawia University, Subratah, Libya*

[2]*Geology Department, National Engineering School, Sfax University, Sfax, Tunisia*

ABSTRACT

Waves are the most important phenomena affecting marine navigation, either in the field of fishing, military or transporting of goods. This paper tries to answer the following important questions: What are the causes and types of waves in the coast of Zuara city? What are their characteristics? And how do those waves affect the marine navigation and human activities on this coast? The research finds significant results devoted in that: the coast is exposed to on type of waves; wind waves. Zuara coast has never been exposed to waves of Tsunami or landslides. The largest size of the wave forms in winter season due to the wind of north-west which is the fastest wind type that the Libyan coast is exposed to. However, the highest speed is up to 65 knots accompanied with waves reach a height of more than 7 meters. The research also classifies the wind speeds that lead to cancelling ships and boats trips that depend on this work is studying the waves in Zuara coast and the relationship with the waves of Libyan coast and Mediterranean sea. Also, it focuses on the effect of waves on boat speed, design, fuel consumption, and other effects.

Keywords: Zuara Coast; Marine Navigation; Wind Waves; Dimensions Waves; Marine Human Activities

1. INTRODUCTION

Waves are the circular movement of water molecules that makes the water surface turbulent and makes floating bodies in vertical persistent movement, which does not stop until the waves calm. On this basis, the waves affect the activity of humans who ride the sea, and this causes increasing in the size of waves, leading in many cases to stop the activity and the cancelling of a lot of cruises for the reason of fear or cut the risk of adventure to challenge the sea waves.

Hence, it can be seen that the most human activities affected by the waves are: marine navigation, transport of goods, passengers on ships, fishing and military activities. All these activities are in need to the nature of waves in order to perform their duties. And therefore, specialists in all mentioned areas, to know about the properties of waves in coats that they use in their activities in terms of navigation and the sizes, causes and types of the waves, as well as the time that waves occur, so that they can avoid dangers which sometimes lead to lack of human lives [1].

And because of the Coast of Zuara city has shipping activity which is distinct from the neighboring coast that contains a port for cargo and fishing, so the waves should be studied deeply to find out their nature, size and other characteristics so can be dealt with in a proper way. This paper focuses on the properties of the waves, on their nature and impacts on human activities.

Zuara coast is located to the north-west of Libya longitude 12°04'45E and latitude 32°56'3N, limits Zuara to the north Mediterranean Sea (**Figure 1**), and west of the Libyan capital city (Tripoli) about 100 km.

However, the study also examines the relationship of waves with the maritime navigation in the coast of Zuara city and tries to answer the following questions:

Q1: What are the causes and types of waves?

Q2: What are the properties of waves in the coast of Zuara city?

Q3: How do waves affect marine navigation?

On this basis, some objectives appear which are: studying the causes and types of waves, knowing the nature of the waves in terms of dimensions and times to find out the impact of waves on marine navigation.

The researcher has followed several methods and approaches to reach the objectives of the study, which are the statistical method for dealing with meteorological data, the descriptive approach and field visits.

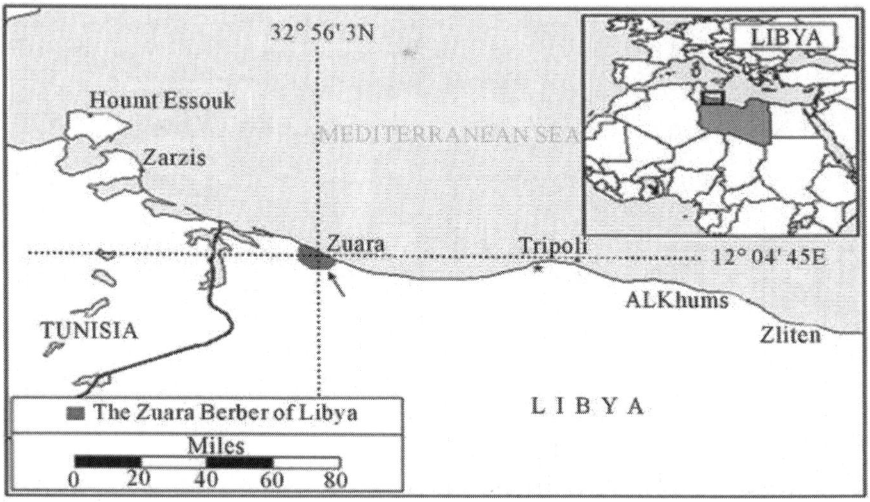

Figure 1. Zuara city locations [2].

This study is an initial step to open search in this kind of few studies in the country, and this is what induces me to investigate in this area of research.

2. THE CONCEPT OF WAVE AND ITS DIMENSIONS AND TYPES

Wave is defined as the regular height (convexity) and depression (concavity) in the water surface. This is called a waved relating to the wave [3].

On this basis, a wave has several dimensions as shown in **Figure 2**:

A—Bottom of the wave (trough), which is the lowest point in the concavity posed by the depression of water surface.

B—The top of the wave (crest), which is the highest point in the convexity posed by the rising water.

C—Wave height: the distance between the bottom of the wave and its peak, because the wave falls below sea level and rises the same distance depression.

D—Wave length: the distance between two crests or two troughs and there is an inverse relationship between wave height and length that as length increases the height decrease.

E—Wave amplitude, a widening distance between the overall level of water surface in a calm sea, and the crest of the wave. This is important especially in the case of specifying the height of the industrial jetties in ports.

F—Wave period: the time it takes for peaks to pass a fixed point [4].

G—Wave steepness is the percentage of wave height to the wave length H/L, and in deep water the wave does not contain the decline greater than 1/7 without a breakage, but in some times waves may decline slightly bigger than this value. Waves can't have appointed top angle more than 120° (**Figure 3**) without breaking [5].

3. THE WAVES CAUSES

Waves happen by several causes or factors, namely:

1) Landslides, mudslides or the collapse of dams;
2) Earth quakes and volcanoes that happen at sea and produce the so-called tsunami;
3) Winds, storms and hurricanes.

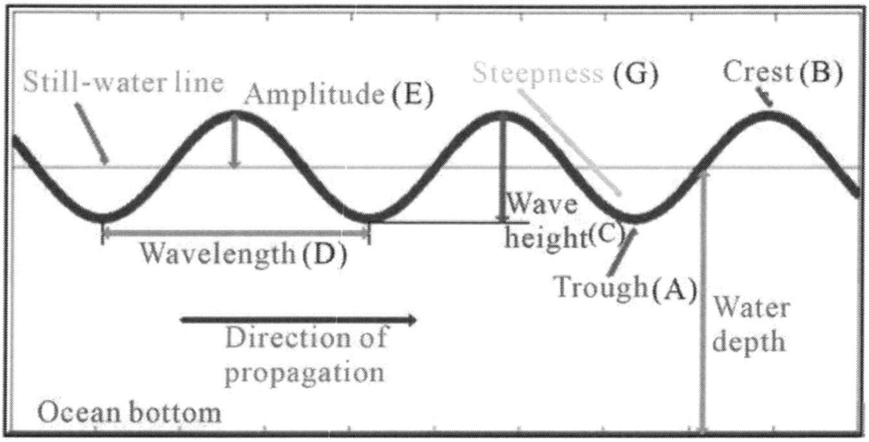

Figure 2. Dimensions of a wave [6].

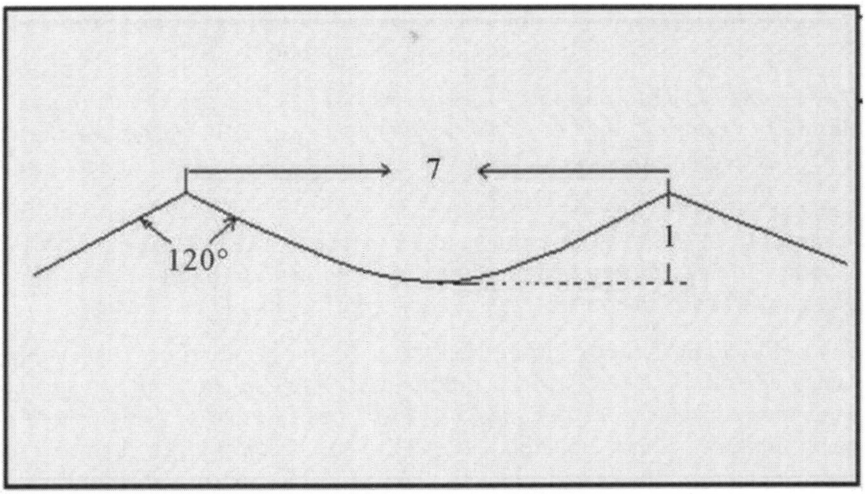

Figure 3. Wave can not have a pointed top or decline of more than 7/1, without break.

3.1. Landslides, Mud Slides and the Collapse of Dams

Landslides, which happen in the relatively small water bodies cause destructive waves. On 9 October 1963, for example—Italy has seen the catastrophe collapse of huge clay blocks in the basin of (Fayont) dam, which led to the formation of waves 90 meters high in outrageous release to a distance of 1.5 kilometers towards the lower valley devastating everything in its way [7].

But such waves are rare to occur and non-observable only if they occur in relatively small lakes or rivers so they will not get the detailed attention, especially as their occurrence in the sea is ineffective.

3.2. The Seismic Waves (Tsunami)

Earth is subject to about one million earthquakes a year. People mostly do not feel them because either they are weak or occur in unpopulated areas. Usually human feel the earthquake only when its magnitude up to 4 on the Richter scale, and is considered a major earthquake when its strength is more than 7 degrees on this scale.

When marine earthquakes occur, sea water dominated by giant waves, which called the "Tsunami". Tsunami is a Japanese word originally means the tidal waves, but these waves are not related to tidal, so specialists call it

"Seismic waves" relative to the earthquakes [8]. Also tsunami means port waves (perhaps because the ports are more susceptible to these waves).

These waves are built up in the deep sea water, and attack the coast pour about 100 thousand tons of water on every square meter of the beach and thus lead to greater losses than the damage caused by the earthquake itself.

Tsunami waves are generated by a volcanic eruption, earthquake on the ocean bottom surface, landslides in the bottom, or a land collision with cosmic body generating a violent jolt. These all lead to an offset section of the sea water which creates a turbulence changing into giant waves.

There are changes occurring in the bottom of the seas and oceans, leading to movement of the land surface and the emergence of cracks in the bottom of the sea caused by earthquakes and volcanoes (**Figure 4**). When there is displacement or land subsidence in part of the ocean bottom surface, this leads to format an empty space that quickly filled with water, and as a result of quick flow of water toward the bottom to fill the space, this makes the water flows away from the coast, and according to fluid mechanics engineering to bounce in reverse direction of flow of water, where a rush of large amounts of water from the ocean bottom surface upward as a result of back reaction on the water speed. These amounts of water surging up causing a big increase in the level of water surface in the middle of the ocean, taking off from this level driving waves towards the coast. These waves moving too fast, and the wave speed depends on the strength of the earthquake its situation from the sea bottom. Its destructive power strength lies in that it suddenly appear to people on the beach, rising abruptly from 1.5 meters to approximately 40 meters and hits the coast with tremendous force could cover the land to several kilometers.

The tsunami waves have the ability to pass the entire ocean and also have the ability to maintain its full power during the journey, and this is where the danger lies that

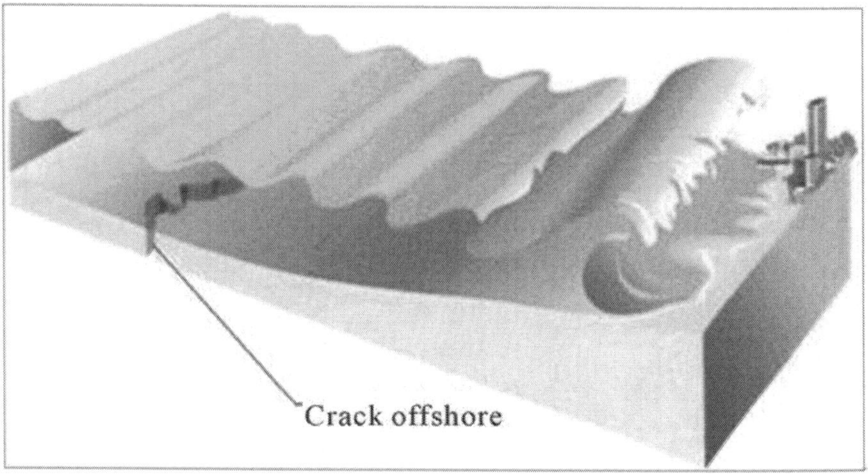

Figure 4. Crack in the bottom caused by the earthquake and offshore [9].

the wave energy does not decrease as it moves toward the beach, but maintain its level; because of the flexibility in energy transfer through the water molecules as observed in **Figure 5**, which shows the movement of the tsunami from the earthquake's center to of most the Indian Ocean coasts passing thousands of kilometers to hit the beach in a full capacity, leading to heavy losses in lives and properties.

3.3. Geographic Distribution of Tsunami in the Mediterranean Sea

When observing the earthquake on the earth map it is found that it is not distributed at random, but concentrated mostly in the major belts stretching long distances across continents and seas, and most of those belts powerful are the "Ring of Fire", extending along the east coast of the Pacific, and makes a belt along the west coast of the Pacific [11].

But in the Mediterranean Sea, belt goes a long across Africa, Europe and Asia. This belt is the most closing belts to the Libyan coast that extends from Morocco to Tunisia through Algeria. And from Tunisia extending north across the Mediterranean to Italy, Greece and Turkey, even China, and this belt is known as the belt of the Alps and contains about 21% of the world's earthquakes [12].

But earthquakes happen in places which have no relation to earthquake belts mentioned above. When the earthquakes are strong, they have serious consequences as the earthquakes that hit Turkey from time to time and in

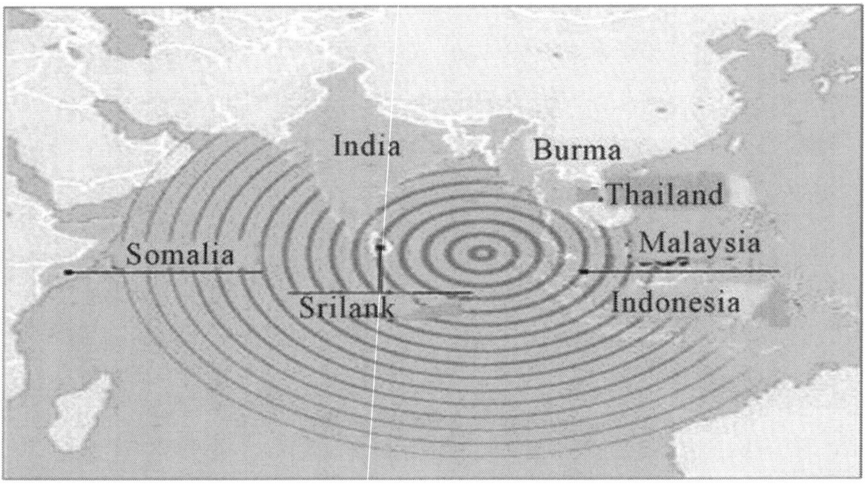

Figure 5. The tsunami transition between the Indian Ocean is full destructive power [10].

Algeria and elsewhere. Most of these earthquakes are not included tsunami for several reasons: the small size of the Mediterranean compared to the tsunami space requirements that give the opportunity to consist largely and destructive, as well as the basin containment of islands that working on cracking the waves, also shallower water in most of coastal areas. All of these factors preventing the occurrence of the devastating tsunami waves that hit the coasts of the Indian Ocean and the Pacific. However, tsunami was recorded in the basin several cases such as: the 29 February 1960 earthquake of Agadir (Mw 5.9) was the most damaging event in Morocco. The 10 October 1980 of El Asnam (Ageria) was the largest (Mw 7.3) recorded in North Africa and revealed 36 km of surface thrust faulting. The 21 May 2003 Zemmori earthquake (Mw 6.8) affected the area immediately in the east of the capital Algiers that experienced an average 0.50 m coastal uplift. The 1994 earthquake sequence struck the Rif region in Morocco and exhibited seismic.

In the Libyan coast until now researcher found no scientific evidence of benefit exposure of the Libyan coast tsunami in spite of the presence of foci of earthquakes, especially in the eastern region near the coast of Green Mountain. See **Figure 6**.

Figure 6 shows the coast of Libya divide into three main seismic focus areas, zone (III) represents the west coast of Tripoli in the east of the Tunisian border. And zone (I) is the coast of Sirt Gulf and stretching from Benghazi in the east to Khomess in the west. Zone (II) is the coast of Green Mountain.

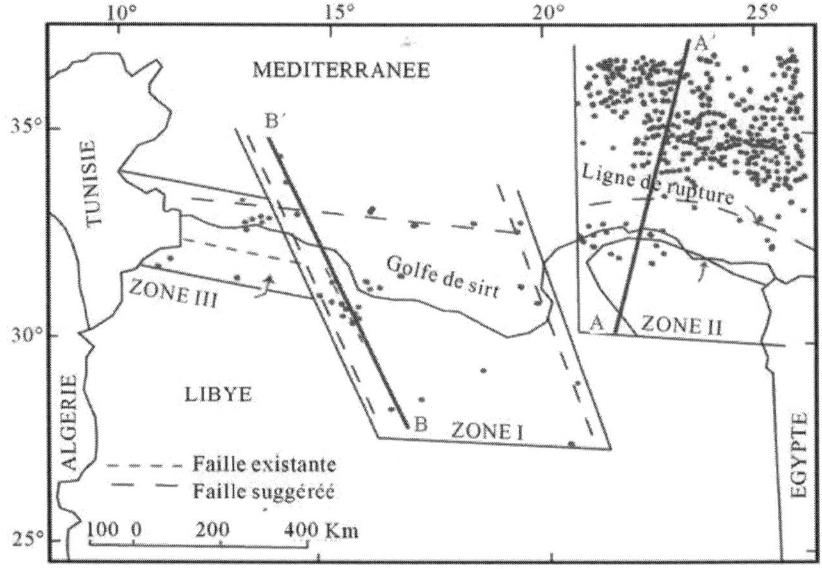

Figure 6. Distribution of seismic fuci in Libya [13].

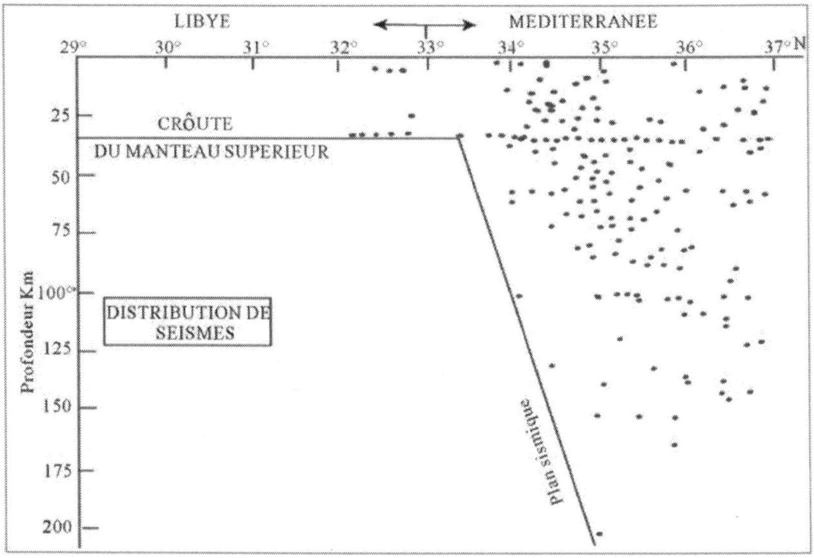

Figure 7. A vertical distribution of seismic foci in the region (II) [14]. The Coast of Green Mountain.

Intensifying spots in zone (II) at the coast of Green Mountain in **Figure 7** explains the section head of coastal zone (II). Notice a number of heavy seismic infesta tions, the vast majority of them in the sea, and the rest few (9 spots) on the mountain, and is characterized by focus mountain proximity to the surface where it does not exceed its depth from the surface of the mountain 40 km. The seismic focus located in the sea spread over different distance from the Earth's surface to a depth of 200 km.

In fact, despite the existence of a huge amount of seismic spots at the bottom of the Green Mountain coast, tsunami is not registered, and this does not mean it doesn't exist, and may for these waves have been monitored for the lack of attention to these phenomena, because they are not affecting directly on people lives in this region.

From **Figure 8** zone (I), the area includes the Gulf of Sirt stretching from Benghazi east and even Misrata in the west, at least in this region the focus of earthquakes introduces **Figure 8** distribution and few outposts in the sea (3 spots) at a depth approximately of 30 km, and increase in land (9 spots) at a depth ranging between 20 and 40 km, and this means less likely exposed to tsunami. The study area (Zuara coast): there is no any seismic focis, as to hotbeds coastline from Tripoli east to Ras Igedir west. See **Figure 6**, which makes us confident that no exposure to tsunamis.

4. WIND WAVES

This type occurs almost every day in the water masses, except for static air days, which is rare, especially in a coast like Zuara coast, which is located within a transitional unstable climate. The size of waves depends on wind speed and the duration of blowing and other factors.

How Wind Waves Are Brew

When air pressure differs between two adjacent areas, the air travels from the region of high pressure to the other of low pressure. When it moves above the water surface, broken short capillary waves occur in at first by the direct contact between air molecules and the surface of water, just as the impact of blowing in a pot of water where wrinkles in the surface of water occur.

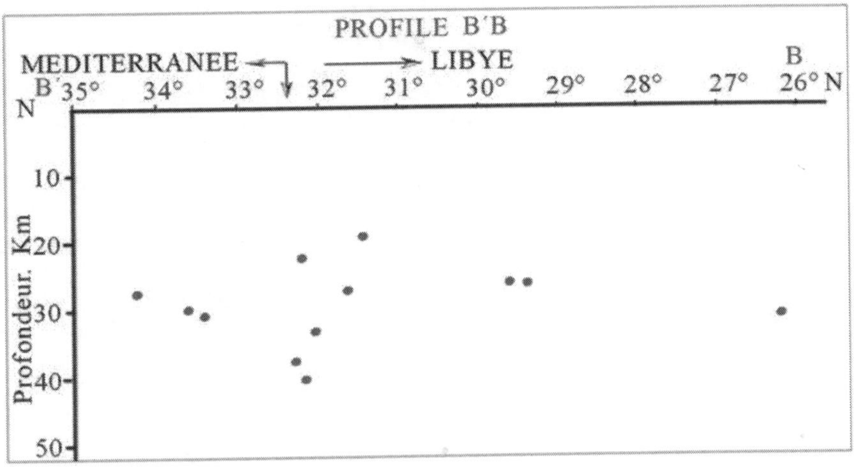

Figure 8. The vertical distribution of the seismic foci in region (II) [15].

When an expert fisherman's in a small boat feels in the movement of wind and observes formation of small broken waves, he quickly goes back to the safe harbor. Because he knows that these small waves will be transformed into the bigger waves and the sea will become more eruptive.

When small capillary waves break, they build larger waves as they transport wind power to the water and as the period of wind gets longer, it gives the chance to the bigger waves to be formed and increases the length and speed of the waves to become equal to the speed of wind.

From **Figure 9** it can be observed that the incurrence of wave top to wind directly helps to push orbits of water molecules forward, while the swirling air in the bottom of the wave causes intensive turning of molecules backward.

The expansion of water body or the distance that the wind is blowing (Fetch) also affects the wave size and speed, where **Figure 10** shows how the smaller waves give the opportunity for waves to be greater as fetch increases.

In addition, to the period of wind and water extent, the volume and speed of waves also affected by wind speed; so there is a direct proportional that as the wind speed increases an increase of wave size and speed happens, as it is concluded from **Table 1**.

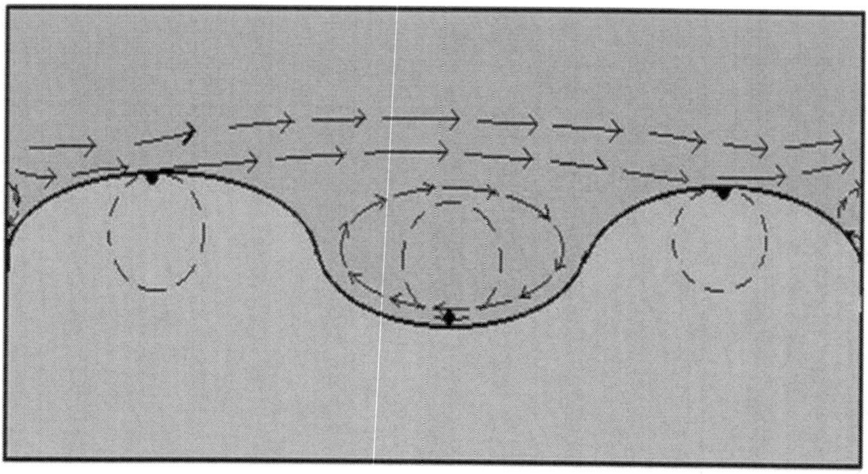

Figure 9. The role of wind for wave generation.

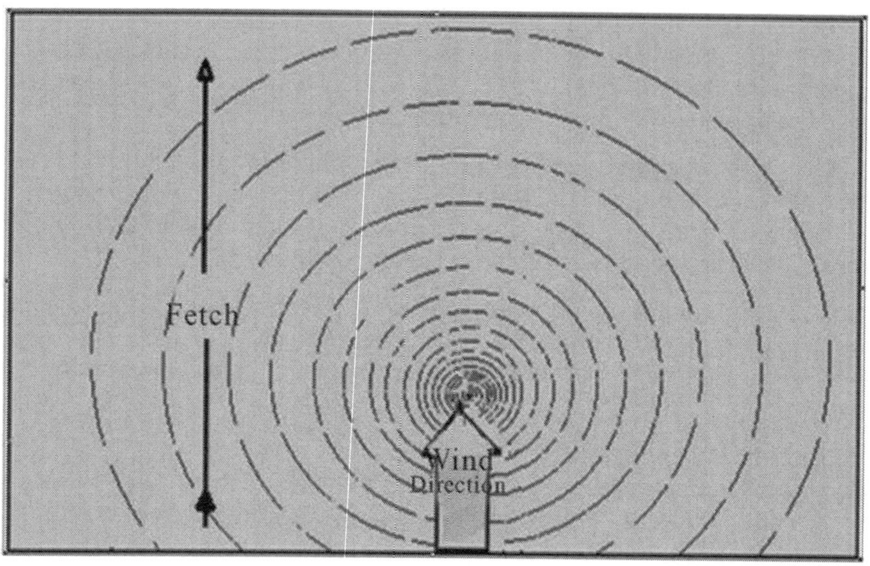

Figure 10. An increase in the size and speed of waves with the increased breadth of the flat fetch.

Therefore, some specialists tried to determine the relationship among the wind speed, sea appearance and wave hight. One of those specialists is the British Admiral Sir Beaufort Franz in the British navy in 1806.

Through the previous view of wave types, it is clear that the coast of Zuara and the Libyan coast in general is exposed only to waves caused by wind, so we should give a general idea to the kinds of wind experienced by the study area.

5. THE WIND IN THE YEAR SEASONS

5.1. The Wind in the Winter

Due to the instability of weather in this season, for the reason of passing many air depressions, it is found that this region is nearly exposed to all kinds of wind. But these kinds differ in the percentage of recurrence, and the most prevailing winds in winter are the south-west wind, followed by southern winds, which blowing in this region at the front of depressions and characterized by severe coldness, particularly in the early morning. That is because it is coming from cold region, and then coming the north, north-west and west winds which blow in the rear of depressions these winds are often rainy on the coast. In addition, to the mentioned winds other wind types reach the coast such as: north-east and south-east and all these winds are dry.

Table 1. The effect of wind speed and fetch on the length and height and wave period [16].

Wind distance (Fetch)	Wind speed (Km/hr)	Length of wave	Period of wind	Wave height/m	Wave period
20	20	2.4	0.3	0.4	4
30	60	6.0	0.9	1.1	6
40	140	10	1.6	2.6	8
50	300	16	3.0	4.5	10
60	500	23	4.5	7.0	12
75	1200	42	9.0	14	16
90	2600	69	15.0	25	20

5.2. The Wind in the Summer

This season is characterized by the stability of weather and clear climate as well as lack of rainfall and due to the centralization of an area of low pressure on the Sahara between the temperature rise in this season and high atmospheric pressure on the north of the Mediterranean Sea, the coming wind in this case mostly from the north and north-east. This means, from high pressure area to low pressure area, and the most prevailing wind is the north-east and north wind, respectively. This dry wind prevails all parts of this region and is working for the temperament of temperature for the reason that it is coming from lower temperature area as east winds. In addition, other wind types are rarely blowing as south east and Western winds.

5.2.1. Wind in the Spring and Autumn Seasons

The wind blows in these two seasons from all directions, but the most blowing winds come from the north east and east south. In this case the weather is characterized by its instability.

The most dangerous wind types on population is the wind blows in the end of spring season and the beginning summer season as well as in autumn sometimes. This wind is called “Gibly”, which is a local nomination. This wind is characterized by its sever dry and high temperature, and in many cases is loaded with dust, as often calm at sunset and return next day, and may continue for several consecutive days in the same month. In March which considered the most month exposed to this type of wind, where the percentage of blowing reached 17.2% over 21 years [17]. The reason for blowing of this wind type is the depression on the Atlantic Ocean, which rushes to the East of the Mediterranean through the Strait of Gibraltar where Gibly wind gravitate to it, since its direction starting with south east then to south and eventually south west following the direction of depression coming from west in the east direction.

From the previous, it is clear that this coast area (study area) is exposed to the following wind types: 1) Northeast trade winds; 2) Southern winds (Gibly); 3) Trade north and north-western cold winds; 4) Breeze of sea and land.

5.2.2. North East Winds

This kind of winds prevails in summer and working to lower the temperature, but this kind causes the waves to buildup. The wind speed is usually slow in the morning and gradually increases to reach a maximum in the evening, so the fishermen take advantage of early morning in fishing operations.

5.2.3. Southern Winds

Southern winds are local winds blowing from the desert in the end of spring and the beginning of summer, as well as autumn season. When this wind kind blows, it causes a sudden rise in temperature as well as rapidly lowers the relative humidity. Gibly brings the blasting heat and dust storms from the heart of desert, accompanying by an increase of sea water temperature. It also works to calm the waves and still the sea water remarkably, to be a better atmosphere for trips and fishing.

5.2.4. North and North-Western Winds

This wind blows during winter, and in fewer cases during spring and autumn, as well as in front and back of the atmospheric depression invading the Mediterranean from west to east, causing a drop in temperature and rainfall. This wind sometimes brings polar winds causing the Perishing storm. It works to reduce the temperature of sea water to its lowest levels in winter. Also this type of wind causes a buildup of waves since this region is experienced to the highest and most serious waves on the maritime navigation in this season, leading to a decline in the number of trips sailing and fishing and lack of production and experiencing noticeable prices to rise during the winter months.

5.2.5. Sea and Land Breeze

During the day, the sun heats land, leading to heat the air which expands and rises to the top and pressure decreases. While on the sea the atmospheric pressure rises as a result of cooler water relative to the land, and so the air moving from sea to land is called the sea breeze. But at night the opposite occurs, since the land temperature drops less than water where air moving from land to the sea (land breeze). The following **Table 2** shows the dates of sea breeze and land breeze in the area of study. The influence does not exceed one kilometer at the most and makes the strip adjacent to the beach area colder than the south areas, therefore, making this area as a haven for residents in the hot days.

6. WIND WAVES IN ZUARA

The wind speed varies from time to time and from one region to another. In 1905 Beaufort classified wind by its speed and its relationship to waves into twelve classes. It starts from calm state (zero degree) and reaches the hurricane (Class 12). From Beaufort scale shown in **Table 3**, it can be noticed that waves grow higher and risk increases speed of 6 km per hour have light ripples without foam with wind speed. The waves that come with wind or crests, and the wind of speed 85 km per hour accompa nying high types in the form of chains collapsing on each other and impair vision.

Table 2. The time of sea and land breeze [18].

Time	Case of wind
From sunrise to 9:00	Stillness
12:00	Beginning of the sea breeze
14:00	Top of the sea breeze
18:00	End of the sea breeze
20:00	Beginning of the land breeze
5:00	Summit land breeze
Sunrise	End of the land breeze

Table 3 shows the highest speed of wind as relating to forming the highest waves. It is noticed that the highest speed Zuara city coast exposed to in the thirty years is 65 knots. This speed leads to the formation of high waves, which cancel all cruises for both ships and boats. These waves make a major threat to ships cruising at sea, but this region was exposed to this speed only once during the thirty years in November 1987. Dangerous speeds are found but less severe than the speed mentioned and frequently happen. These speeds are 40 to 50 knots which are repeated every three to four years, but recurrence speeds are 20 to 30 knots.

This leads to the cancellation trips, especially the small, medium and large fishing boats which are the most important boat types.

7. THE EFFECT OF WAVES ON THE HUMAN ACTIVITIES

Waves are the most common types of surface marine movement that affect the sea human activities, either directly as in the field of fisheries and maritime navigation, or indirectly, as the positive impact of waves on the marine environment by providing sea with oxygen and cleaning of contaminants. The following is some clarifications these effects:

Table 3. The highest wind speed in knots for the period from 1970 to 2000 [19].

Year	Months											
	1	2	3	4	5	6	7	8	9	10	11	12
1970	24.0	24.0	30.0	35.0	25.0	32.0	27.0	30.0	30.0	30.0	21.0	30.0
1971	32.0	36.0	33.0	30.0	27.0	27.0	30.0	27.0	30.0	30.0	30.0	32.0
1972	29.0	30.0	30.0	30.0	35.0	27.0	25.0	26.0	28.0	25.0	25.0	28.0
1973	27.0	25.0	30.0	27.0	24.0	25.0	27.0	30.0	21.0	25.0	19.0	25.0
1974	30.0	45.0	33.0	30.0	27.0	21.0	18.0	20.0	23.0	35.0	18.0	32.0
1975	27.0	35.0	30.0	28.0	30.0	25.0	22.0	28.0	22.0	25.0	20.0	25.0
1976	45.0	30.0	37.0	27.0	27.0	27.0	30.0	30.0	35.0	40.0	40.0	27.0
1977	25.0	30.0	33.0	35.0	27.0	28.0	30.0	25.0	30.0	28.0	20.0	30.0
1978	50.0	35.0	35.0	30.0	30.0	32.0	25.0	28.0	26.0	25.0	28.0	25.0
1979	35.0	30.0	35.0	35.0	35.0	28.0	28.0	28.0	30.0	35.0	35.0	30.0
1980	33.0	38.0	23.0	38.0	36.0	30.0	28.0	30.0	28.0	40.0	42.0	45.0
1981	45.0	35.0	40.0	40.0	35.0	29.0	30.0	28.0	25.0	30.0	36.0	27.0
1982	38.0	38.0	40.0	35.0	30.0	29.0	22.0	20.0	20.0	35.0	28.0	28.0
1983	30.0	28.0	40.0	26.0	42.0	25.0	20.0	25.0	25.0	24.0	22.0	30.0
1984	22.0	30.0	23.0	25.0	28.0	30.0	22.0	25.0	23.0	28.0	28.0	22.0
1985	38.0	25.0	25.0	35.0	25.0	20.0	20.0	23.0	22.0	28.0	25.0	35.0
1986	23.0	24.0	28.0	25.0	25.0	28.0	28.0	20.0	30.0	30.0	25.0	35.0
1987	23.0	25.0	23.0	20.0	33.0	13.0	33.0	19.0	18.0	28.0	65.0	20.0
1988	30.0	28.0	25.0	30.0	25.0	30.0	28.0	21.0	20.0	25.0	33.0	35.0
1989	21.0	32.0	30.0	25.0	30.0	30.0	20.0	21.0	22.0	29.0	18.0	28.0
1990	32.0	21.0	28.0	30.0	35.0	27.0	25.0	24.0	18.0	20.0	35.0	35.0
1991	25.0	20.0	29.0	30.0	25.0	30.0	20.0	20.0	20.0	25.0	23.0	32.0
1992	19.0	25.0	30.0	25.0	28.0	26.0	25.0	18.0	18.0	20.0	24.0	22.0
1993	22.0	27.0	22.0	30.0	25.0	27.0	20.0	18.0	24.0	22.0	20.0	27.0
1994	35.0	27.0	28.0	35.0	21.0	20.0	21.0	20.0	20.0	20.0	25.0	23.0
1995	25.0	18.0	22.0	32.0	18.0	25.0	20.0	21.0	21.0	23.0	21.0	19.0
1996	17.0	30.0	22.0	27.0	21.0	17.0	25.0	20.0	28.0	35.0	27.0	35.0
1997	20.0	25.0	25.0	39.0	27.0	20.0	25.0	19.0	30.0	18.0	28.0	30.0
1998	29	20	37	29	24	20	20	20	25	25	20	25
1999	28	24	25	25	20	28	23	20	20	19	20	30
2000	20	25	21	24	23	20	22	20	20	25	15	23

7.1. The Impact on Marine Navigation

The navigation is the most human activities affected by waves as they deal with it directly. This effect of waves is on several nautical things, including:

7.1.1. Impact on the Cruise Possibility

Through interviews with some navigation experts from fishermen in Zuara port for fishing and crew they said that there is a specific wind speed that sailing can not be done with it, because this speed will lead to the formation of high waves which make boats and ships to sink, but the larger size of boat the less risk of sinking. The navigation trips are cancelled in wind speeds [20], as follows:

1) Ships

a) Cancel the trips of ships with size of 200 meters, if wind speed increases above 50 knots;

b) Cancel trips of ships with size of 100 meters, if wind speed increases above 35knots;

c) Cancel trips of ships with size of less than 100 meters, if wind speed increases above 30 knots.

2) Boats

a) Cancel trips of boats with size of 20 to 30 meters, if wind speed increases above 25 knots;

b) Cancel trips of boats with size of less than 20 meters, if wind speed increases above 20 knots;

c) Cancel trips of boats with size of less than 15 meters, if wind speed increases above 15 knots.

7.1.2. Impact on the Speed of Ships and Boats

The impact of waves is not only to cancel trips, but on the speed of boats. High waves may allow making trips, but boats will take twice the time that should take when do not face waves, and this means the boat sailing in the opposite direction of the waves. On the other hand, boats may shorten the time of travel when sailing in the same direction of wave movement. This issue needs to be carefully examined to determine the time of travel, taking into account the direction and speed of waves.

7.1.3. Impact on the Fuel Consumption

There is no doubt that the resistance of waves to boat body, when sailing in the opposite direction of waves, requires more power to push it forward. This power needs to increase the amount of fuel, to the extent that the boat journey may consume twice the amount of fuel consumed by the same trip, when it is moving in the same direction of waves.

7.2. The Impact of Waves on the Ports and Harbors

The most important thing in studying waves is to know the maximum height that waves can reach. In the studying area, waves should not be higher than bumpers to avoid extreme impact on the boats lying in those ports and harbors. So, studying waves can help to determine the height of quay or harbor.

7.3. The Impact of Waves on Ship Design

Ship design is similar to building design in terms of size that the strongest earthquakes can not break small kennels. On the other hand, long buildings should be flexible as earthquakes or storms can not break or overturn them. Small boats should be built as solid as possible and big ships should be flexible to withstand shocks and sprains due to weight.

In contrast, while it is commonly believed that the high raging sea waves are non-destructive to ships, long waves compared with the length of the ships are suffering from circular movement of water molecules, similar to cork suffering, and modern ships can withstand small needed bending and twisting, that are built on the basis of flexibility. Mostly, metal sheets on the hulls of ships are built in interference with each other as shells found on the body of fish that it can slide on each other, and leak water. This water is often used in cooling the engine and loads.

One of the things destructing ships is the broken waves. In areas of storms, where waves get bigger and broken as well, water that slides to the front of wave, does not move in circular movement, but it may considered within the amount of water which moves at a high speed in the direction of ship.

This amount of water can strongly destroy parts of ships, which lie above the main surface of the ship or break the hull, especially if strongly pushed to the side of the ship which is above the water surface. So the captain will direct the ship to the reverse direction of wind in the storm. This is because the bow of ship can withstand the wave shocks better than any other parts in the ship. But away from the center of the storm it does not matter in any direction the ship goes, that there is no broken waves and islands. From the other point of view, the study of waves is also important in ship design, where a large amount of energy for pushing ship goes in forming waves. Waves which are behind the ship represents the lost energy from the ship, where the energy is transferred away to be stored at the end, when waves is broken on the distant shores. Thus, the key in designing an efficient boat is to make a body which leaves the smallest impact possible at the economic speed and this can be in the bow, front shoulder, back shoulder and steering wheel. The idea is to get a bigger destructive overlap possible within the waves generated from these points, and thus ships leave the smallest impact possible. So the less energy gives the effect, the more efficient ship.

7.4. Reduction of Wave Activity

In times gone fishermen have realized that they can reduce rough seas on their boats by slowly oil leaking above the water, and the interpretation due to the property of surface tension in liquids, which makes water surface coherent. For example, the mosquito stands on the water surface, and when oil is put on the water doing the action of a rubber plate water surface surrounding the boat, which helps to prevent wave from breaking by the same speed it breaks up as it should.

In addition to this, waves can be calm by putting anything else in water which helps to undermine the wave energy. Ice or waste floating on the surface can do this, and also mud or sea grass can help. In fact, the artificial kelp is made in order to help making waves more calm in places, where the seaweed does not live, but all these solutions are done in a limited area especially boats mooring areas or the descent or ascent of their passengers.

8. CONCLUSIONS

8.1. The Results of the Study

1) There are several types of waves as they occur, and the most important waves are waves of wind and hurricanes, waves of earthquakes and volcanoes, and waves of landslides.
2) The Mediterranean basin is exposed to earthquakes on most coasts and infrequent to accompany tsunami waves such as Algeria and Morocco coasts.
3) Most of the earthquakes do not include tsunami for several reasons: the small size of the Mediterranean compared to the tsunami requirements of space that give the opportunity to consist large and destructive, as well as the basin contains islands that work on cracking the waves, also shallower water in most coastal areas.
4) The Libyan coast is exposed to earthquakes except Zuara coast and no attestation for tsunami waves.
5) Waves in Zuara coast are of the wind wave type.
6) The size and speed of waves in Zuara coast depend on the wind speed and direction, and its blowing time.

7) The largest size of waves is formed in winter by the cause of north-west winds, which the fastest winds experienced by the Libyan coast.

8) The best days of navigation and fishing are the days of south wind where the sea is calm.

9) Zuara coast is exposed to wind speeds somewhat dangerous because they tend to stop the trips of ships and fishing boats. The highest wave speed reached in Zuara coast in 30 years is 65 knots.

10) Zuara coast is the best area for maritime navigation along the Libyan coast, because it is protected from the north-west wind by the Tunisia coast, which is advanced northward from the west, as well as shallow water and long distance, which makes the waves to be broken far from the coast and makes the coast quiet, compared with the east and middle coast of Libya.

8.2. Recommendations

1) Creation of a modern observatory equipped with marine monitoring devices, whereas the current meteorological station is not designed to monitor the marine weather events. Thus fishermen in western Libya depend on the marine bulletins broadcast by the media in Tunisia.

2) Providing free meteorological data to the researchers, and not withheld from them, and the abolition of fees imposed on the researchers for these data.

3) This research is an initial step to study waves and the results are preliminary, and it should be followed by more steps under the guidance of students to study such topic.

4) The waves in Zuara coast have the smallest danger along the Libyan coast and it should take the advantage of this feature and support the fishing and commercial activity by and paying attention to Zuara port from its size and the fleet size working in it.

5) Arranging for courses in marine meteorology for all fishermen and managers of marinas, and fishing licenses should be granted only after attending these courses.

REFERENCES

1. J. O. Algraeo, "Physical Properties of the Waters of the Libyan Coast," Unpublished, p. 100.
2. www.google.com/imgres?q=Zuara+MAP
3. J. O. Algraeo, "Physical Properties of the Waters of the Libyan Coast," Unpublished, p. 88.
4. J. O. Algraeo, "Physical Properties of the Waters of the Libyan Coast," Unpublished, p. 4.
5. A. Lefevre, "Seas and Oceans," Champers World Library, Glasgow, 2004.
6. http://cdip.ucsd.edu/?nav=documents&xitem=waves#irregular
7. K. Isitso, "Oceangraphy," The National Library in Baghdad, 1986, p. 492.
8. K. Isitso, "Oceangraphy," The National Library in Baghdad, 1986, p. 493.
9. http://ts2.mm.bing.net/th?id=H.4558876844296177&pid=1.7&w=227&h=112&c=7&rs=1
10. www.Abonawaf.com
11. E. Almquili, "Hydro Geomorphological Risks," Dar Shomoa al Takafa, Azzawia, 2003, pp. 82-83.
12. J. O. Algraeo, "Physical Properties of the Waters of the Libyan Coast," Unpublished, p. 4.
13. A. J. jusqu'à, "Geology of Libya," 1977.
14. Campbell, et al., "The Geology of Sirt Basin," 1968, p. 659.
15. Campbell, et al., "The Geology of Sirt Basin," 1968, p. 658.
16. K. Isitso, "Oceangraphy," The National Library in Baghdad, 1986, p. 521
17. K. Isitso, "Oceangraphy," The National Library in Baghdad, 1986, p. 523.
18. Libyan NAtional Meteorological Center-LNMC.
19. Central Department of Meteorology/Tripoli
20. A Personal Interview with a Shipmaster Nadm Mustafa Naji, 6 April 2012.

2

CHAPTER

Finite Element Numerical Method for Nonlinear Interaction Response Analysis of Offshore Jacket Affected by Environment Marine Forces

Seyyed Mahmood Ghassemi Zadeh[1*], Reza Shojayee Baghdar[2], Seyyed Mohammad Saleh Vaziri Kang Olia[3]

[1]*Marine Structures Engineering, Amirkabir University of Technology, Tehran, Iran*

[2]*Hydraulic Structures Engineering, Shahrood University, Shahrood, Iran*

[3]*Structure Engineering, Ferdowsi University, Mashhad, Iran*

ABSTRACT

In this paper a nonlinear response of a fixed offshore platform under the combined forces of waves, wind and sea currents is presented. Wave force acting on the elements is calculated using Morison equation. Hydrodynamic loads on horizontal and vertical tubular members and the dynamic response of offshore fixed platform coupled with distribution of displacement, axial force, and bending moment along the base of the platform for regular and severe cases have been investigated. The structure must be able maintain production in a one-year wave return period condition and also to be able to continue with one hundred-year storm return period. The results of this study show that bending moment values with a one-year wave return period condition for the base platform and junction of platform to deck are 70 percent and 59 percent, respectively more than bending moment with a one-year wave return period. The direction of wave and wind hit has significant effects on the shift platform response, also nonlinear response is important for the safe design and operation of offshore structures.

Keywords: Fixed Jacket Platform, Finite Element Method, Stokes Wave Theory, Nonlinear Response of Structure

1. INTRODUCTION

The total number of offshore platforms in the Gulfs and oceans throughout the worlds is increasing each year. Most of these platforms are of jacket platforms types installed in water depths of 32 meters to 200 meters for exploration of oil and gas. Analysis, design and installation of offshore structures compatible with the environmental conditions are of the most challenging and innovative work in this area. Offshore jacket platforms are typically designed using offshore structures standards such as (API RP2A WSD, 2000), (API RP2A LRFD, 1993) and (ISO 19902, 2007). Nonlinear static analysis, i.e. the Pushover analysis is widely used in offshore standards such as API, ISO and DNV to study the nonlinear behavior and the final capacity of offshore platforms against the sea loads. In this method, the corresponding load pattern evenly increases until the collapse of the jacket platform under marine environment affected by special loads, for example, the (wave return period of 100-years). Reference [1] presented studies on offshore structures under the action of the wave's effects. Reference [2] carried out studies in the field of nonlinear hydrodynamic forces and their effects on the nonlinear response by platforms. Reference [3] presented studies on designing reliable offshore structures under the action of the waves. Reference [4] made studies on the dynamic response of a steel jacket platform with an effective specific activity and control under the wave's load and the impact of these forces on the response of a structure with and without vibration control mechanisms can be simplified. Reference [5] presented the effect of an active control system for a jacket platform under wave loading. Their studies showed that the proposed algorithm has better effect than conventional control strategies. Reference [6] analyzed the time domain of dynamic response of simplified offshore structures under the concurrent effects of accidental loads and seismic wave and showed the results of the displacement response of the time history and value of RMS. Reference [7] studied dynamic response reliability analysis of offshore structures under the effects of marine waves, ocean currents and earthquakes and showed that reliable indicators in long-term behavior and final responses on offshore structures have large effects. Reference [8] [9] conducted studies on the response of jacket platforms with active mass dampers under the force of the waves and showed that the active mass dampers TMD can reduce the respond of platform jacket considerably. Reference [10] [11] showed the effects of seismic estimates on coastal and offshore structures according to the most recent earthquakes and showed the effectiveness of viscoelastic dampers.

In this paper, nonlinear analysis is formulated for reliable assessment of the response of jacket platform under constant structural loads, wave, sea currents and wind loads. A three-dimensional finite element model is used to estimate the displacement and stress in steel platforms under combined constructional and waveloads. This analysis includes a variety of nonlinear properties produced due to changes in the nonlinear drag force. Wave and current flow kinematic are generated by the 5th order stokes nonlinear theory. The effect of wave forces on members is calculated by Morison formula. Natural periods and modes of the system are calculated. Nonlinear wave kinematic is a very important factor due to the interaction between structures and waves. Wave induced loads on fixed offshore platform under the sea storms is calculated using a nonlinear drag term in Morrison equation and changes in wave height. Moreover, the fixed jacket platform response under the ultimate structural loads is a function of the behavior of components in the range of nonlinear deformations.

2. THE ENVIRONMENTAL LOADS

The water force can be classified into waves force and the current flow force. The wind blows over the ocean moves the water causing the current flow and the waves. Ocean waves force on the platforms is dynamic and natural, although the static design of platforms in shallow waters is also acceptable. With increasing waters depth, the platforms are flexible with more dynamic effects.

2.1. Current Flow Forces

Wave leads to the orbital motion of the water. The orbital motion in a closed cycle, but slightly drive forward due to surface wind effect. Current flow is generated by the wave. The current flow in a wave tends to drag wavelength. The induced current flow force on the cylindrical structure is defined as follows [12] :

$$F_D = C_d \frac{\rho}{2} A U^2 \tag{1}$$

In this equation, F_D, C_d, ρ, A, and U, are the drag force in Kilo Newton, drag coefficient, sea water density per ton per cubic meter, cross section depicted in terms of square meter and current flow rate in meters per second, respectively. The flow rate is usually between 0.1 and 2.3 meters per second. Profiles of sea currents in the Gulf of Mexico are shown according to the API [13] in Figure 1.

2.2. Wave Forces

The regular wave theories are used to calculate the forces on fixed offshore structures, illustrated inFigure 2, and based on the three parameters of water depth (d), wave height (H) and wave period (T), respectively. Hydrodynamic force vector is calculated in all degrees of freedom. The intensity of the wave force on each meter of the structure is calculated based on Morison formula shown in Equation (2), [13] .

$$F = \rho C_m V \frac{du}{dt} + \frac{1}{2} \rho C_d A u |u| \tag{2}$$

where, F, C_m, u, du/dt, and V are total force exerted on an object, the coefficient of inertia, current flow rate, current flow acceleration and object size, respectively.

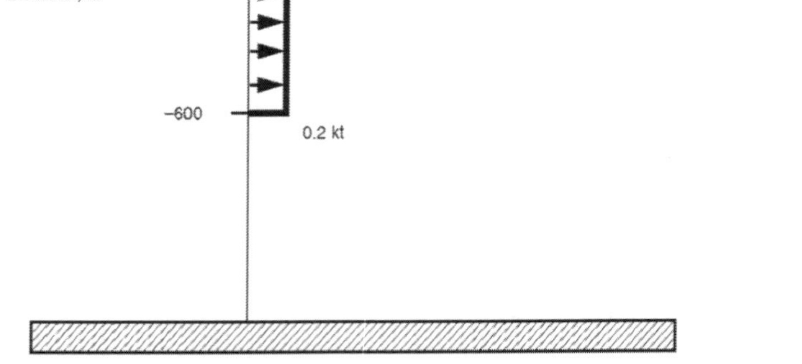

Figure 1. Sea currents profile in the Mexican gulf.

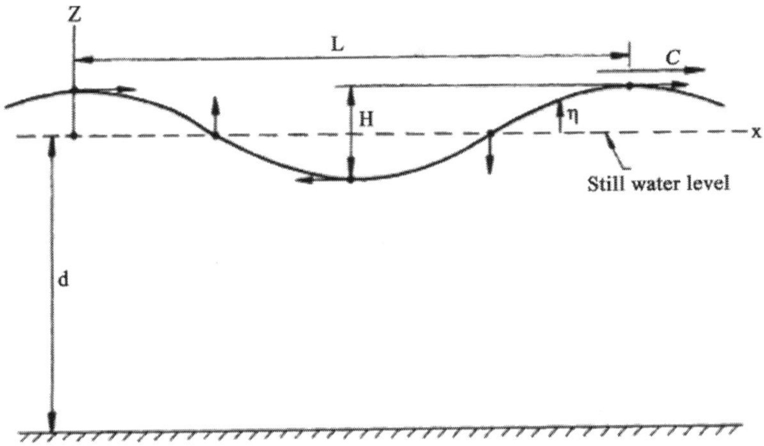

Figure 2. Wave parameters profiles.

2.3. Wind Forces

When a structure is placed in the path of the moving air so that wind is stopped or is deflected from its path all or part of the kinetic energy is transformed into the potential energy pressure. Therefore wind forces on any structure result from the differential pressure caused by the obstruction to the free flow of the wind. These forces are functions of the wind velocity, orientation, area, and shape of the structural elements. Wind forces on a structure are a dynamic problem, but for design purposes, it is sufficient to consider these forces as an equivalent static pressure. The force of the wind according to the API is shown in Equation (3).

$$F = 0.0437 C_s A_p U_{10}^2 \quad (3)$$

In this equation, F, C_s, A_p and U_{10} are the wind force in Kilo Newton, shape structure factor, cross section depicted in terms of square meter and wind speed in kilometers per hours, respectively.

3. Parametric Topics

3.1. The Finite Element Method

The finite element method is for the approximate solution of differential equations governing on the continuous areas. This method was initially proposed as a method of stress analysis and is widely used for this purpose. For a non-persistent problem, finite element approximation is defined as follows [14] :

$$\frac{\partial f}{\partial t}+U\frac{\partial f}{\partial x}=0 \tag{4}$$

By transforming this equation into the integral one, and by using the weighted residual function ($W_{(x)}$), we have:

$$\int_{\Omega}\frac{\partial f}{\partial t}W\mathrm{d}x+\int_{\Omega}U\frac{\partial f}{\partial x}W\mathrm{d}x=0 \tag{5}$$

Finite element is defined as follows:

$$f(x)=\sum\nolimits_{I} f_I N_I(x) \tag{6}$$

Using the relations (5) and (6), this equation will change as follows:

$$\sum_{I}\frac{\partial f_I}{\partial t}\underbrace{\int_{\Omega}N_I N_J\mathrm{d}x}_{M_{IJ}}+U\underbrace{\int_{\Omega}N_I\frac{\mathrm{d}N_J}{\mathrm{d}x}\mathrm{d}x}_{K_{IJ}}=0 \tag{7}$$

In the above equation, M_{IJ} and K_{IJ}, represent the mass matrix and stiffness matrix respectively.

The finite element method uses the integral shapes of equations; it means that integration have to be carried out on governing equations of the scope of problem; Therefore, finite element is only used for discretization the local terms of the equations and for discretization of the time terms the finite difference method is used consequently [14] .

3.2. Nonlinear Wave Theory

In the theory of nonlinear waves, the sinusoidal shapes of wave profile and rotation of the particles in a circular path in the profile does not exist simultaneously. One of the nonlinear wave's theory in 1880 was presented by Stokes; it was based on adding unlimited number of successive approximation and adding it to a series of equations. In this research, the 5th order Stokes nonlinear wave theory is used to estimate the water particle kinematics. In this theory, by using the two characteristics amplitude and period of the wave, water particle kinematics is determined using the following equations. One of the ways by which the higher order Stokes theory can be estimated is Fenton method [15] . The wave profile can be written as:

$$\eta(x,z,t)=\frac{1}{k}\sum_{i=1}^{5}\beta_i \cos i(kx-\omega t) \tag{8}$$

Horizontal and vertical velocities are calculated as follows:

$$u(x,z,t)=\frac{\partial\phi}{\partial x}=c_E+\frac{\sqrt{gk\tanh kh}}{k}x\times\sum_{i=1}^{5} i\alpha_i \cosh ik(z+d)\cos i(kx-\omega t) \tag{9}$$

$$w(x,z,t)=\frac{\partial\phi}{\partial z}=\frac{\sqrt{gk\tanh kh}}{k}x\times\sum_{i=1}^{5} i\alpha_i \sinh ik(z+d)\sin i(kx-\omega t) \tag{10}$$

In these equations k is wave number, ω is angular velocity, f is velocity potential function, C_E is average of steady flow, α_i, A_{ij} and β_{ij} are affiliate factor related to k_d that d is the depth of the water. The coefficients are available in reference number [15] .

4. PLATFORM STRUCTURAL MODEL

The platform studied in this paper is a fixed jacket platform that had been proposed for installation in Venezuela Gulf in 1977 [16] . The platform is composed of 8 bases made of hollow cylindrical steel. Outside diameter of vertical, horizontal and diagonal elements are 1.8, 1 and 0.5 meter, respectively; and the thickness of vertical, horizontal and diagonal elements are 2.5, 1.55 and 1.1 centimeter, respectively. In order to meet the hardness requirements on platform, cross bracing was used on both sides of the bases. Also, platform height is 57.5 meter with sea water density of 1.0252 tons per cubic meter. To simplify the model, platform, all deck loads were model in one story. Steel profile used in this platform has the Specific weight: 7850 (Kg/m^3), Elastic Modulus: 2.1×10^9 (Kg/m^2), Shear modulus: 8.077×10^8 (Kg/m^2), Yield stress: 360 (MPa) and Ultimate tensile stress: 420 (MPa).

5. THE PLATFORM FINITE ELEMENT ANALYSIS

Finite element analysis of platform is performed under various wave, sea currents and wind loads. Structural model is focused on a detailed description of the deformation properties of the column loads. These platform columns are modeled by equivalent beams. For the present analysis, the dead weight of all fixed equipment located on the deck is 7.25 ton per square meter and a live load objects on deck of 0.3 ton per square meter is taken. This platform has been installed in water depth of 54.5 meters. C_d and C_m parameters were considered equal to 0.65 and 1.6, respectively according to API regulation. A one-year and 100-year wave parameters are shown in Table 1 Wind speed of one-year and 100-year of 85 and 100 (Knot) were considered according to API regulation.Figure 3 depicts two-dimensional and three-dimensional view of the platform model. Columns 1 and 2 are considered to analyze the platform. Also, the Nodes a_0, a_1, a_2, a_3 and a_4 are considered to show displacement and stress on the columns.

Marine wave and wind parameters and current flow directions of ±0°, ±45°, ±90°, ±135° and ±180° intended for the analysis of these platforms. 19 different load cases have been shown for analysis at this platform Table 2 [17] . In the Table 2 DL, LL, W_a, W_i and C_u, represent dead load, live load, wave load, wind load and sea current load, respectively.

6. NUMERICAL RESULTS

For more effective and accurate design, a finite element model to estimate the internal forces and columns displacements of offshore platform under structural loads and the waves were evaluated. Structural vertical loads are actually static loads, whereas the lateral wave is in oscillation in the time domain directly linked with the angle of the incident wave. The model used in this research is a steel platform proposed in 1977 to the Gulf of Venezuela. Three-dimensional finite element model of the platform is designed in SAP2000 software. Second hand parts such as ladders, stairs and so forth are not directly modeled and only the weight effects are applied. Z axis in a Cartesian system is in the direction of water depth. Fixed end boundary condition is located at 3.5-meter mud line/seabed along with the bases (depth 57 meter). Natural periods and related vibration modes shapes are analyzed by Eigen values. Table 3 shows 12 vibrating modes of the platform.

Table 1. Wave parameters.

Definitions	Water depth (m)	Wave height (m)	Wave period (s)
Wave with a return period of 1-year of operation	54.5	10	7.2
Wave with a return period of 100-year for safety		16	10.6

Table 2. Loading combinations.

Loading combination	Description	Loading combination	Description
Load Comb 1	DL + LL	Load Comb 11	$DL + LL + (W_a + W_i)_{100yr} + W_d/W_f/C_u\ 0^{\circ}$
Load Comb 2	$DL + LL + (W_a + W_i)_{1yr} + W_d/W_f/C_u\ 0^{\circ}$	Load Comb 12	$DL + LL + (W_a + W_i)_{100yr} + C_u\ 45^{\circ}$
Load Comb 3	$DL + LL + (W_a + W_i)_{1yr} + C_u\ 45^{\circ}$	Load Comb 13	$DL + LL + (W_a + W_i)_{100yr} + C_u\ 90^{\circ}$
Load Comb 4	$DL + LL + (W_a + W_i)_{1yr} + C_u\ 90^{\circ}$	Load Comb 14	$DL + LL + (W_a + W_i)_{100yr} + C_u\ 135^{\circ}$
Load Comb 5	$DL + LL + (W_a + W_i)_{1yr} + C_u\ 135^{\circ}$	Load Comb 15	$DL + LL + (W_a + W_i)_{100yr} + C_u\ 180^{\circ}$
Load Comb 6	$DL + LL + (W_a + W_i)_{1yr} + C_u\ 180^{\circ}$	Load Comb 16	$DL + LL + (W_a + W_i)_{100yr} + W_d/W_f/C_u\ 45^{\circ}$
Load Comb 7	$DL + LL + (W_a + W_i)_{1yr} + W_d/W_f/C_u\ 45^{\circ}$	Load Comb 17	$DL + LL + (W_a + W_i)_{100yr} + W_d/W_f/C_u\ 90^{\circ}$
Load Comb 8	$DL + LL + (W_a + W_i)_{1yr} + W_d/W_f/C_u\ 90^{\circ}$	Load Comb 18	$DL + LL + (W_a + W_i)_{100yr} + W_d/W_f/C_u\ 135^{\circ}$
Load Comb 9	$DL + LL + (W_a + W_i)_{1yr} + W_d/W_f/C_u\ 135^{\circ}$	Load Comb 19	$DL + LL + (W_a + W_i)_{100yr} + W_d/W_f/C_u\ 180^{\circ}$
Load Comb 10	$DL + LL + (W_a + W_i)_{1yr} + W_d/W_f/C_u\ 180^{\circ}$		

Table 3. Natural periods of offshore platform.

Mode No.	Period (s)	Mode No.	Period (s)	Mode No.	Period (s)
1	0.332	5	0.208	9	0.129
2	0.269	6	0.191	10	0.122
3	0.243	7	0.150	11	0.104
4	0.211	8	0.142	12	0.104

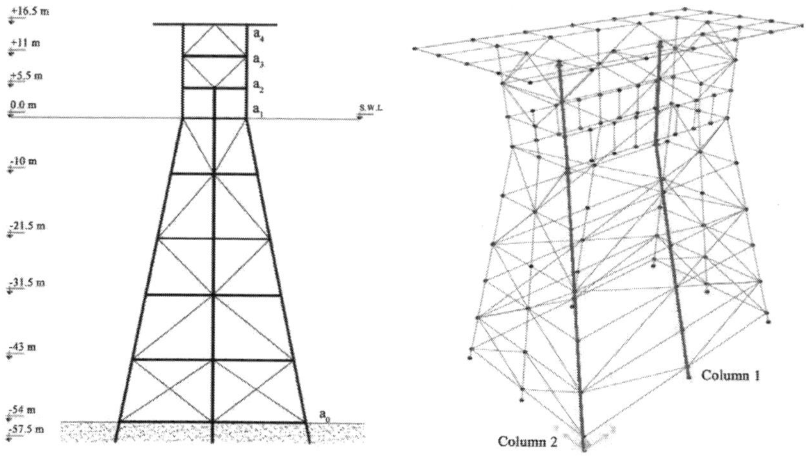

Figure 3. Three-dimensional and two-dimensional view of modeled platform from different views.

6.1. Structure Displacement Response

In order to understand the behavior of fixed offshore platform, the analysis of these platforms in water to a depth of 54.5 meters under the sea waves, winds and sea currents were studied. Maximum deformations under mentioned loads have been precisely calculated. Deformations responses of U_1, U_2 and U_{abs} (which represents the absolute horizontal displacement is equal to the square root of the sum of squares U_1 and U_2) and are depicted in the following figures in line with platform height under the action of waves loads and sea currents with return period of 1-year (operating conditions) and 100-year (extreme conditions/ storm). U_1, U_2 deformation are in X and Y directions, respectively.

For operating conditions of platform, based on Figure 4, Figure 5 and Figure 6, with increasing angle of the sea current attack, angle from zero to 180, platform displacements are reduced, so that the angle of zero degrees shows the maximum displacement and 180 degrees exhibits the minimum displacement in the X direction. This condition is contrary for the Y direction, attacks with angle of 90 degree showing the maximum displacement and the minimum displacement are with zero and 180 degrees angle. Furthermore, the average value of absolute displacement difference for the 1-year sea current for combined number 4 (maximum absolute displacement) and combined number 6 (minimum absolute displacement) is about 11 percent. For operating conditions of platform, based on Figure 7, Figure 8 and Figure 9, with increasing angle of the wave, wind and sea current attack, angle from zero to 180, platform displacements is reduced, so that the angle of zero degrees shows the maximum displacement and 135 degrees exhibits the minimum displacement in the X direction. This condition is contrary for the Y direction, attacks with angle of 90 degree showing the maximum displacement and the minimum displacement is with 180 degree angle. Also the average value of absolute displacement difference for the 1-year wave, wind and sea current for combined number 8 (maximum absolute displacement) and combined number 10 (minimum absolute displacement) is about 16 percent. For storm conditions, as Figure 10, Figure 11 andFigure 12 illustrates, hits with 90 degree angle create the maximum displacement, while hits with 180 degree angle create the minimum displacement. The average value displacement difference for a 1-year wave of combination number 4 (maximum absolute displacement) and combination number 6 (minimum absolute displacement) is at about 8 percent. With increasing angle of the sea current attack, angle from zero to 180, platform displacements is reduced, so that the angle of zero degrees shows the maximum displacement and 180 degrees exhibits the minimum displacement in the X direction. This condition is contrary for the Y direction, attacks with angle of 90 degree showing the maximum displacement and the minimum displacement is with 180 degree angle. Also the average value of absolute displacement difference for the 100-year sea current for combined number 14 (maximum absolute displacement) and combined number 15 (minimum absolute displacement) is about 15 percent. In Figure 13, Figure 14 and Figure 15, with increasing angle of the sea current attack, angle from zero to 180, platform displacements is reduced, so that the angle of zero degrees shows the maximum

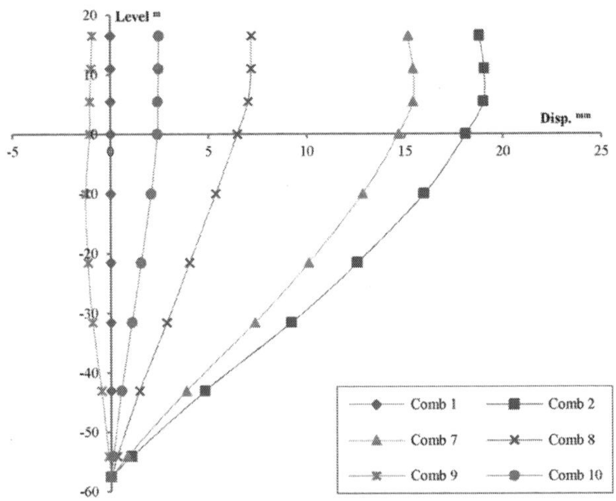

Figure 4. Displacements in the X-direction with 1-year currents Column 1.

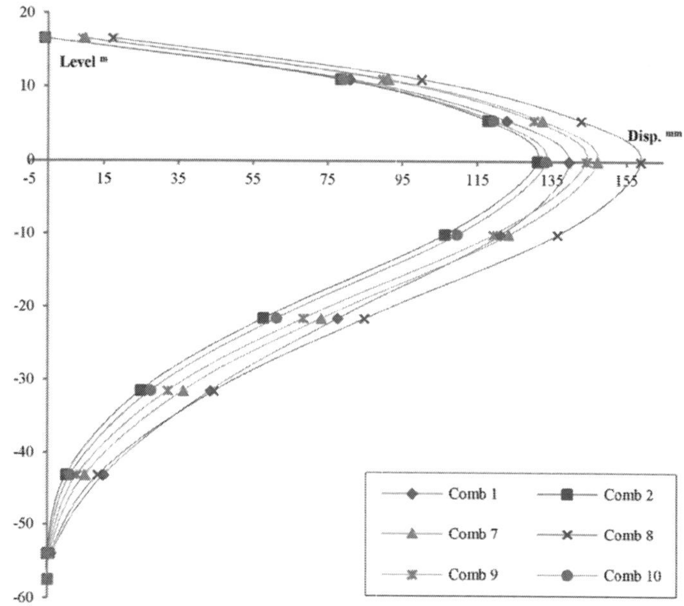

Figure 5. Displacements in the Y-direction with 1-year C_u Column 1.

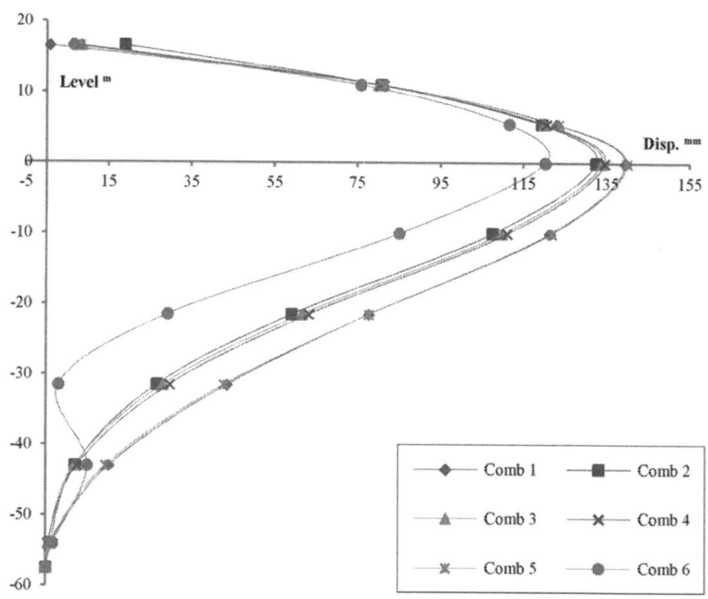

Figure 6. Absolute displacements with a 1-year C_u Column 1.

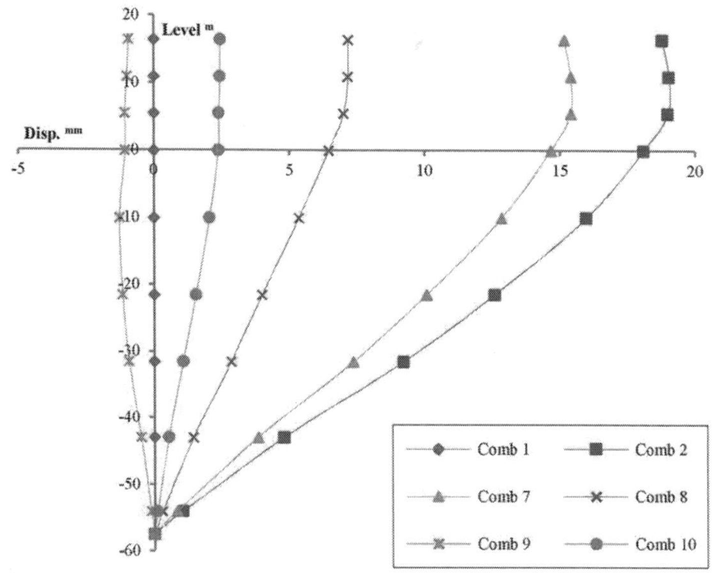

Figure 7. Displacement in the X-direction with 1-year W_a, W_i & C_uColumn 1.

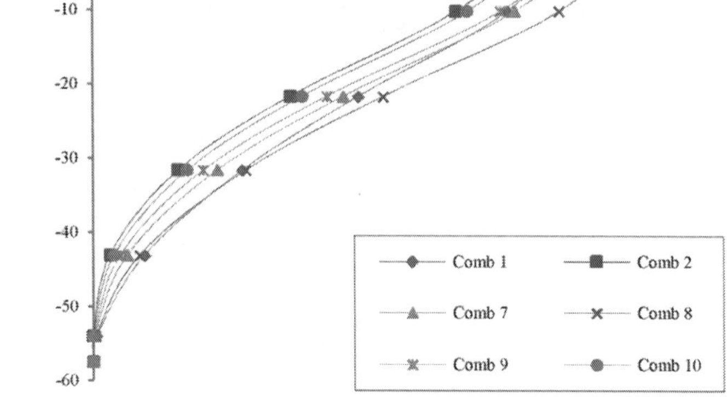

Figure 8. Displacement in the Y-direction with 1-year W_a, W_i & C_uColumn 1.

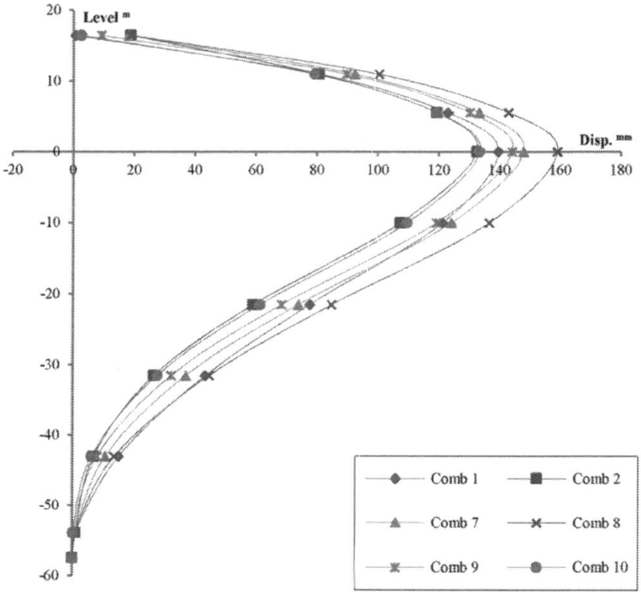

Figure 9. Absolute displacement with 1-year W_a, W_i & C_u Column 1.

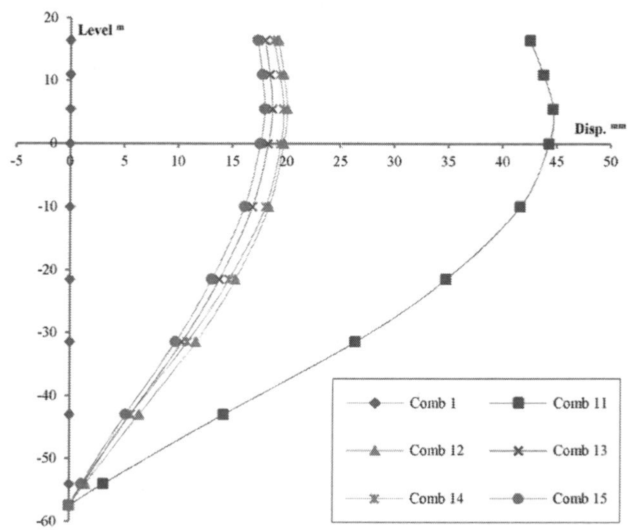

Figure 10. Displacement in the X-direction with 100-year C_u Column 1.

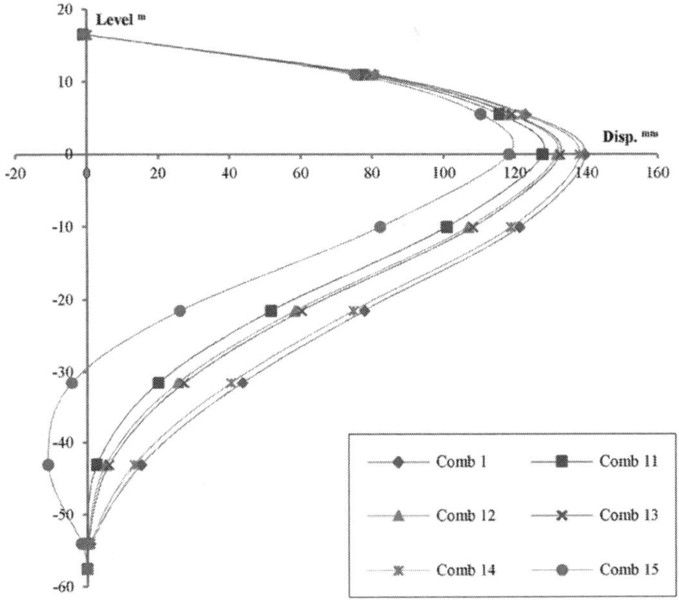

Figure 11. Displacement in the Y-direction with 100-year C_u Column 1.

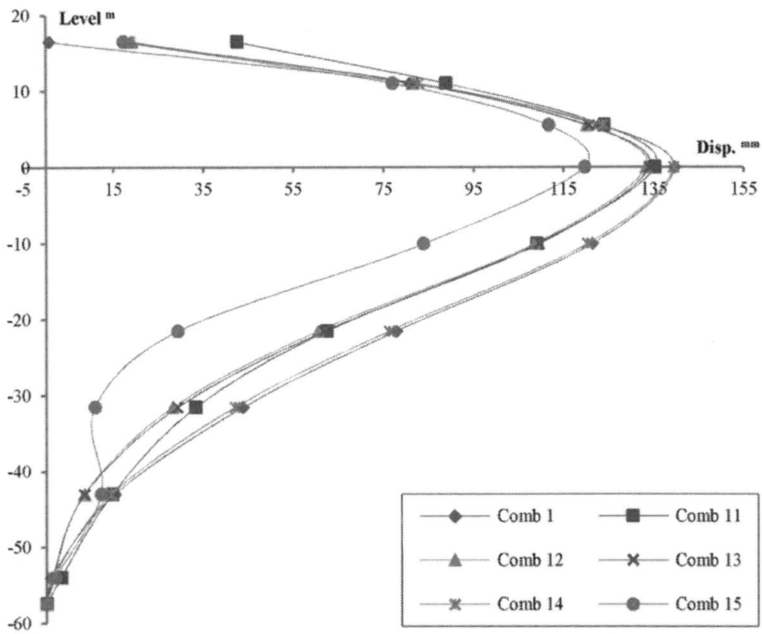

Figure 12. Absolute displacement with a 100-year C_u Column 1.

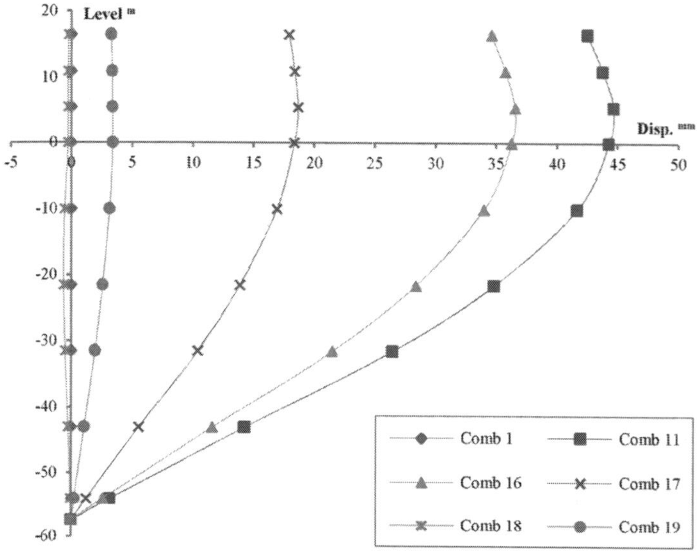

Figure 13. Displacement in the X-direction with 100-year W_a, W_i & C_uColumn 1.

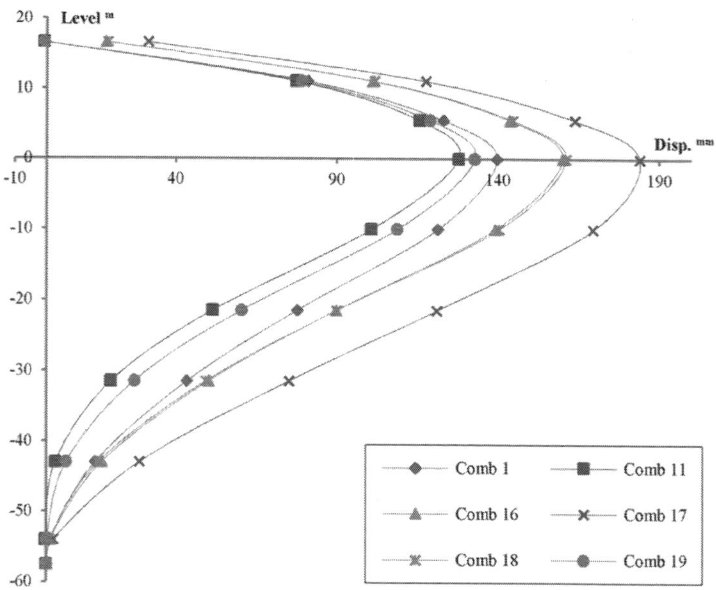

Figure 14. Displacement in the Y-direction with 100-year W_a, W_i & C_uColumn 1.

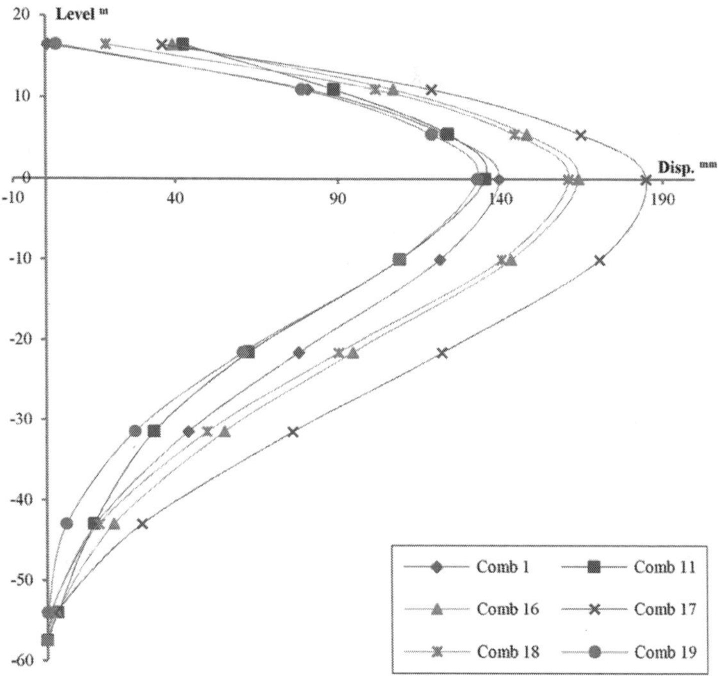

Figure 15. Absolute displacement with a 100-year W_a, W_i & C_u Column 1.

displacement and 135 degrees exhibits the minimum displacement in the X direction. This condition is contrary for the Y direction, attacks with angle of 90 degree showing the maximum displacement and the minimum displacement is with zero degree angle. Also the average value of absolute displacement difference for the 100-year sea current for combined number 17 (maximum absolute displacement) and combined number 19 (minimum absolute displacement) is about 37 percent.

Figure 16 and Figure 17, indicates that the maximum displacement at a_1 (still water level) and a_4(junction between deck and jacket) are under the combined effect of all the loads. As the figure suggests, by changing the operating mode to a storm condition, column 1 displacement becomes more. The average displacement value for the node a_1 in two modes of one and 100-years with the same wave, wind and sea current hit, angles is about 7 percent and for the node a_4 this difference is about 13 percent.

6.2. Axial Forces and Bending Moments Responses

Figure 18, Figure 19 and Figure 20 show the bending moment $M_{2\text{-}2}$, $M_{3\text{-}3}$ and absolute for levels of a_0 to a_3. The values for the bending moments of a_1 to a_3 levels are almost uniformly and changes in hit angles of the wave, wind and sea currents have little effect on the bending. The average bending moment value of one and 100-years with the same wave, wind and sea current hit, angles is about 54 percent. In Figure 21, the maximum amount of bending moment at the base of columns of jacket is shown. According to this figure, in both operating and storm condition, with increase angle of attack of the wave, wind and sea current, the bending moment decrease.

Figure 22 shows the maximum axial force at critical levels in line with the height of the jacket platform structure. It is important in the design of a platform bases that the maximum axial force be considered, because the thickness of the base platform can be reduced with respect to the maximum stresses.

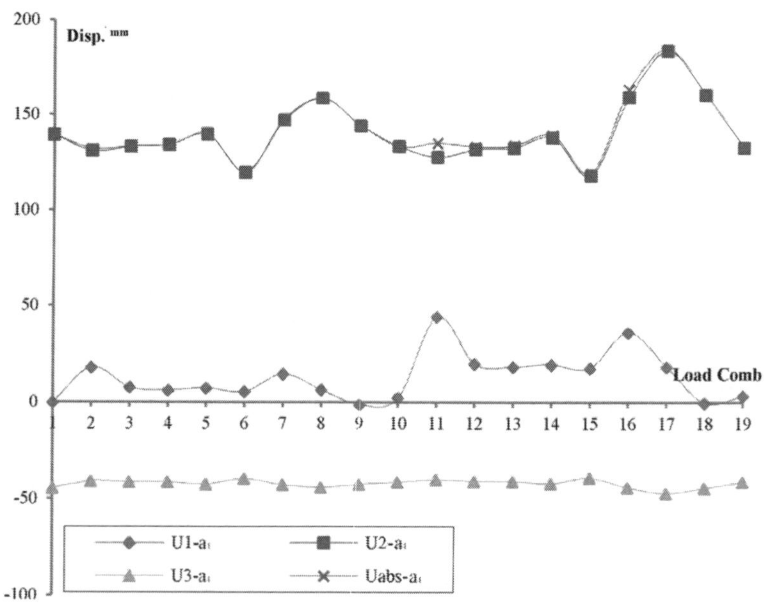

Figure 16. Column 1 displacement varies for a_1 level.

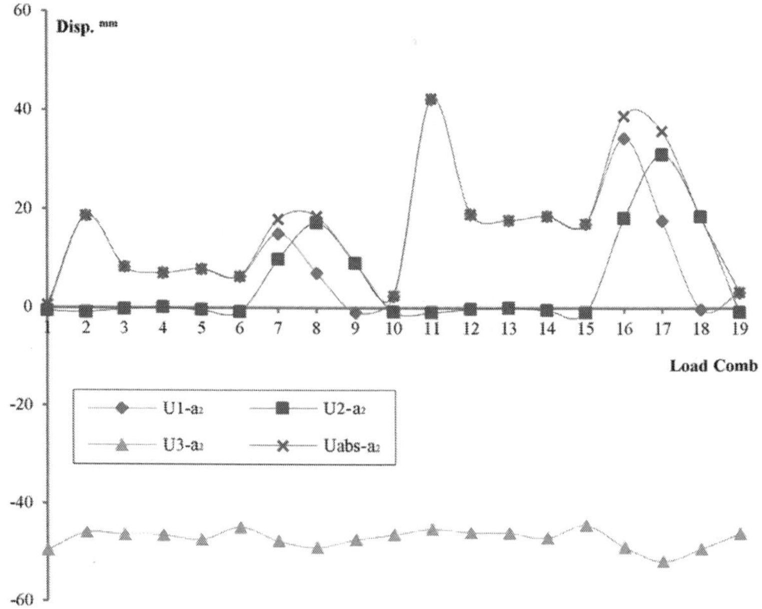

Figure 17. Column 1 displacement varies for a_2 level.

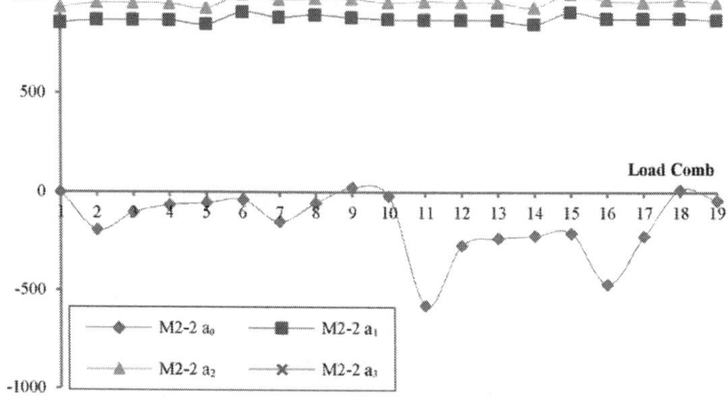

Figure 18. Column 1 bending moment M_{2-2} for different levels.

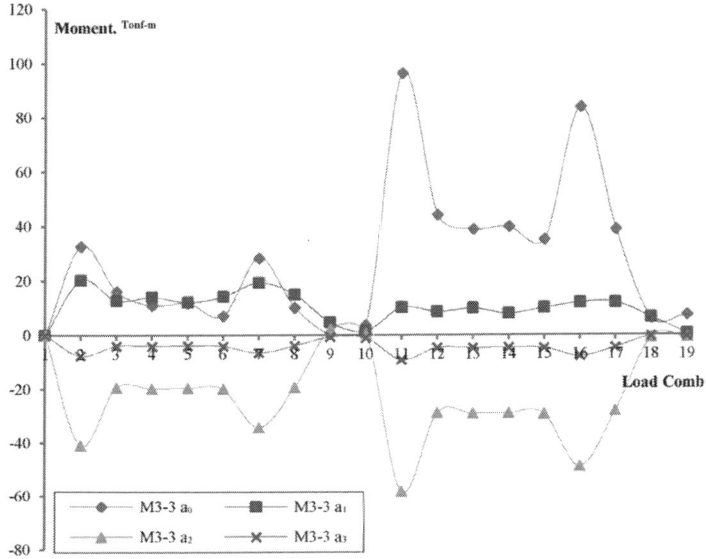

Figure 19. Column 1 bending moment $M_{3\text{-}3}$ for different levels.

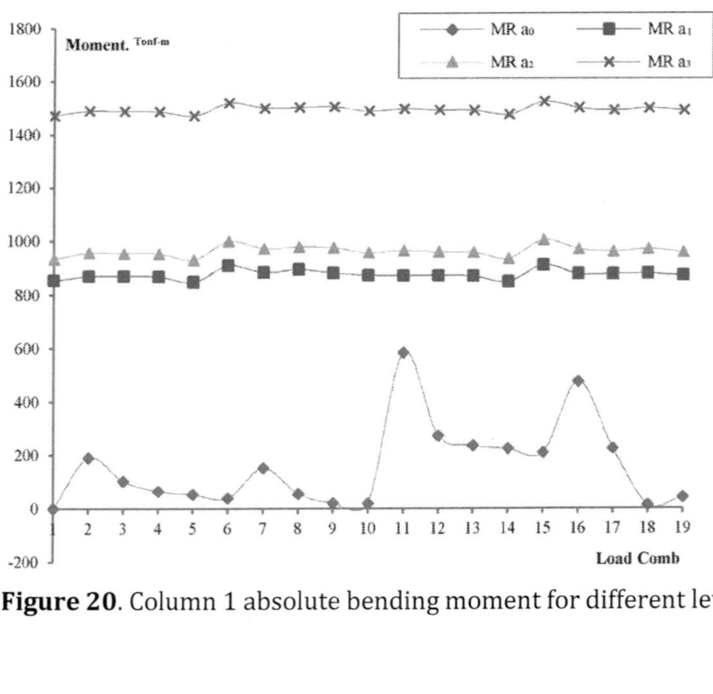

Figure 20. Column 1 absolute bending moment for different levels.

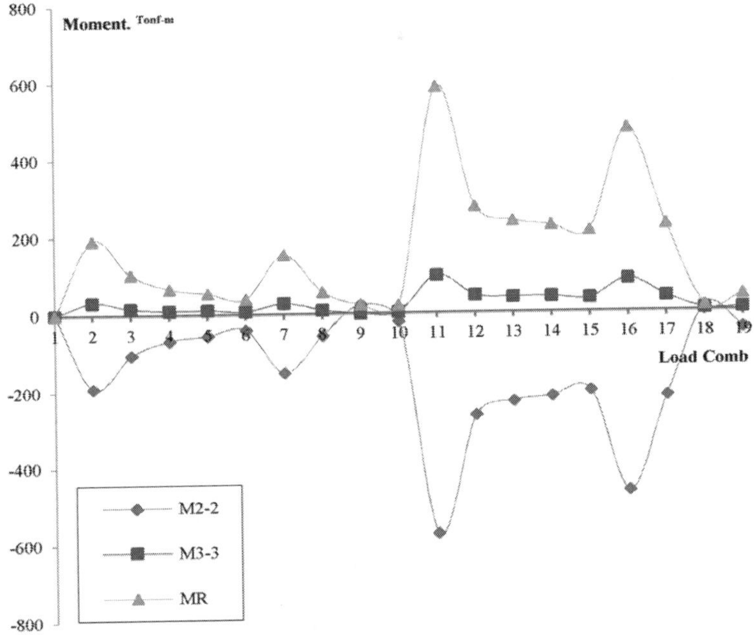

Figure 21. Base of Column 1 bending moment respond.

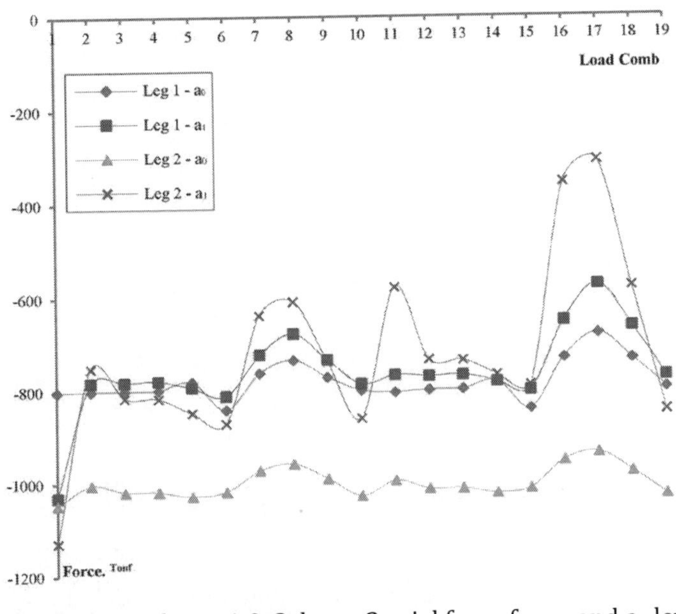

Figure 22. Column 1 & Column 2 axial force for a_0 and a_1 levels.

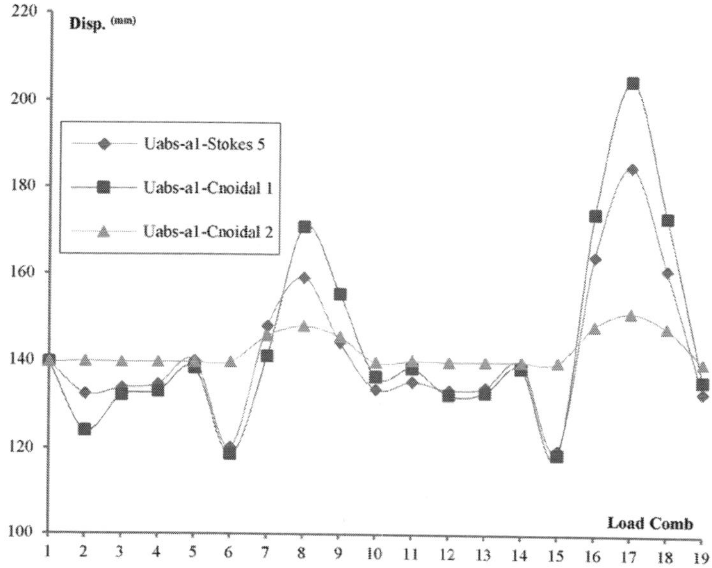

Figure 23. Comparison of displacement with all load combinations Column 1.

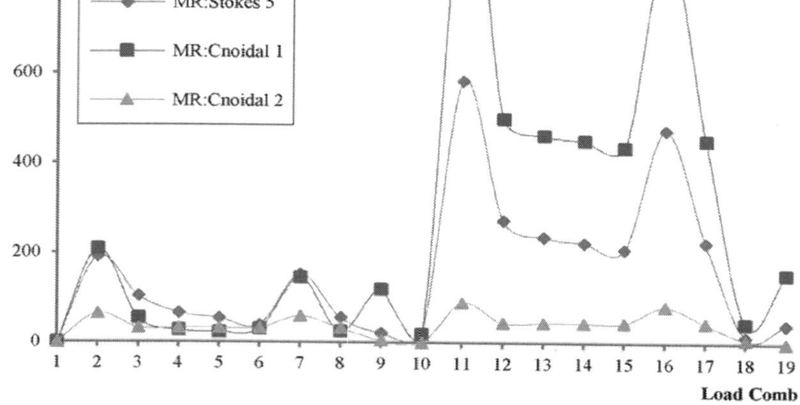

Figure 24. Comparison of moment with all load combinations in Column 1.

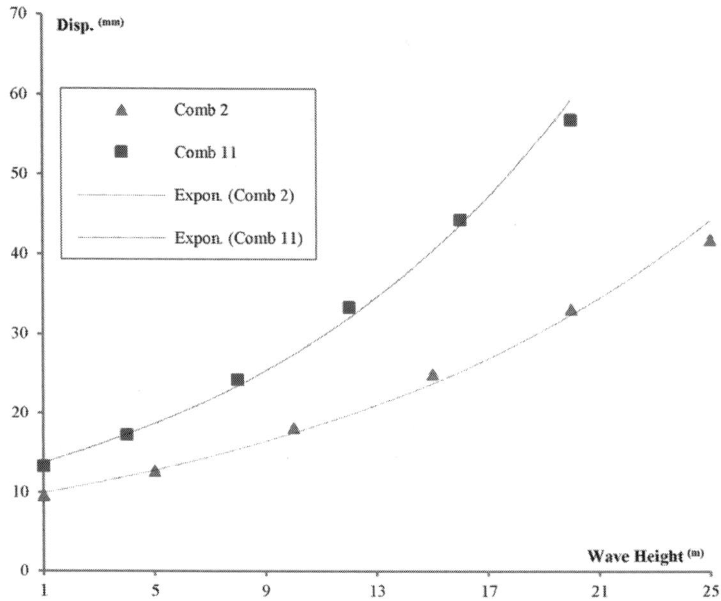

Figure 25. Comparison of displacement with different wave heights in Column 1.

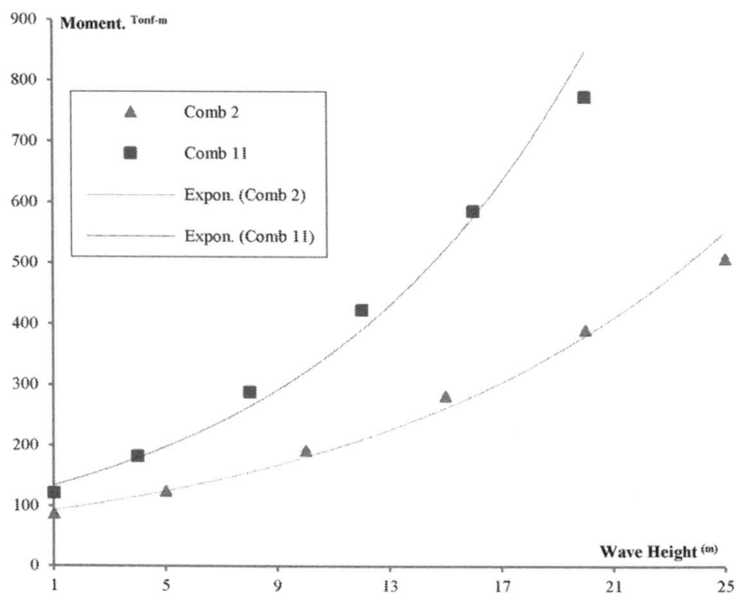

Figure 26. Comparison of Moment with different wave heights in Column 1.

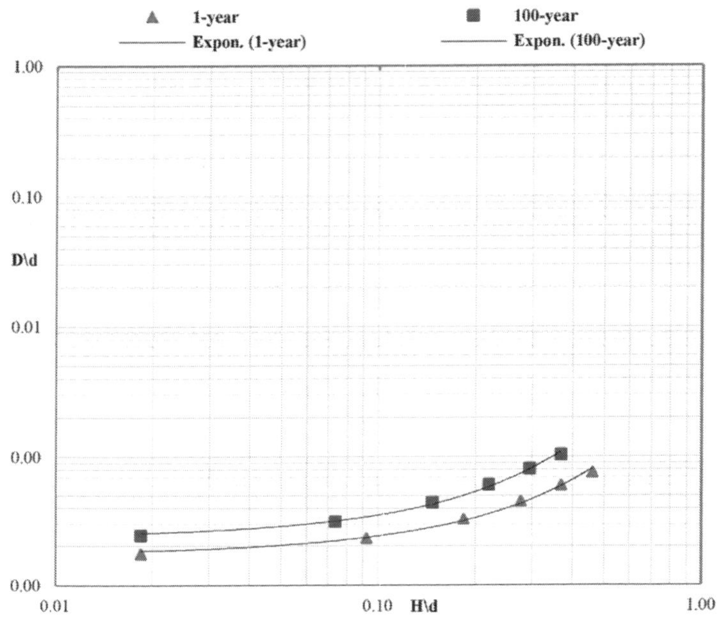

Figure 27. Normalized comparison of wave height on node a_1.

6.3. Parametric Compares

In order to compare the behavior of fixed offshore platform, three different wave theories (Stokes IV, Cnoidal I & Cnoidal II) were developed. In Figure 23 and Figure 24 displacements of (Node a_1) and moment of (Node a_0) under these three different wave theories with all load combinations were shown respectively. As was shown in Figure 23, the average of absolute displacement in column 1 (Node a_1) by using Cnoidal I wave theory is 3.4 percent more than Stokes IV wave theory and 12.6 percent more than Cnoidal II wave theory. In Figure 24, the average amount of absolute moment in column 1 (Node a_0) by using Cnoidal I wave theory is 11.8 percent more than Stokes IV wave theory and 24.5 psercent more than Cnoidal II wave theory.

As respects, load combination 2 shows the maximum displacement and moment with return period of 1-year (operating conditions) and load combination 11 shows the maximum displacement and moment with 100-year (extreme conditions/storm), the parametric comparison on (Node a_1) shows that by increasing wave height, the moment and displacement of jacket increases exponentially. These results are shown in Figure 25 and Figure 26 respectively. In Figure 27, the normalized comparison of wave height on (Node a_1) shows that by increasing (H/d) in which (H) is wave height and (d) is constant water level, the amount of (D/d) in which (D) is displacement of jacket is increases exponentially and can performs by $Y = me^{nX}$ equation, that m is (0.0002) and b is between (3.38 - 4.2).

7. CONCLUSION

Designing effective and affordable offshore platforms largely depends on the proper evaluation of the responses of hit during the useful life of the structures. However, the performance of the platform in various operations in bad weather requires that the entire structure is designed to meet the final condition. The design depends on the site of the platform. It is important that the response of the offshore platform reduce according to environmental loads. In general, offshore structures dynamic stress range reduction to about 15 percent leads to double increase of the service life and thus reduced maintenance costs. Periodic and regular inspections of offshore platforms to issue the certificates assurances require studying the structural response according to wave forces. In this study a finite element formulation is developed to study the nonlinear response of offshore fixed platform. A three dimensional element including large-scale displacements and time dependent wave force as a drag component of the wave force, which is a function of second-order water particle velocity was developed. Structural offshore analysis has been conducted in order to obtain a platforms shift response under different loads. Deformation of platform under combined waves, wind and ocean current flow loads was investigated. Offshore platform displacements, axial forces bending moments and free vibration frequencies were evaluated. The maximum displacement of all nodal points for wave and ocean currents with different angles of incidence was analyzed. The results shows that different angles of sea currents have little impact on the response of the horizontal displacements; while the wave hit directions shows significant effects on the value of displacements response. Displacements response U_1 increases nonlinearly with increasing platform height, but a significant curvature in displacements response U_2 is observed in the height of the platform. The results show that the wave-current flow direction shows little effect on the wave bending moment in a one-year return period, while sea current flow impact direction has a significant effect on the amount and direction of the bending moment. The wave bending moment $M_{3\text{-}3}$ with a return period of 100-years for the a_0 (the base of the platform) and a_1 (junction platform deck) are respectively 70 percent and 59 percent higher than the wave bending moment with a return period of 1-year. Compression between three different wave theories shows that Cnoidal I wave theory made the larger displacement and moment in the jacket than Stokes IV and Cnoidal II wave theories. Also by increasing wave height, the amount of displacement can perform by $Y = ae^{bX}$ equation, that a is between (9.3 - 12.7) and b is between (0.06 - 0.078) and moment can performs by $Y = ce^{dX}$equation, that c is between (86 - 122) and d is between (0.074 - 0.096) increases exponentially.

REFERENCES

1. Hallam, M.G., Heaf, N.J. and Wootton, L.R. (1978) Dynamics of Marine Structures: Methods of Calculating the Dynamic Response of Fixed Structures Subject to Wave and Current Action. Construction Industry Research and Information Association (CIRIA), Underwater Engineering Group, Report UR8, London, 326 p.
2. Chakarabarti, S.K. (1987) Hydrodynamics of Offshore Structures. Springer, Berlin.
3. Barltrop, N.D. and Adams, A.J. (1991) Dynamics of Fixed Marine Structures. 3rd Edition, Marine Technology Directorate Limited, Epsom.
4. Abdel-Rohman, M. (1996) Structural Control of Steel Jacket Platform. Structural Engineering and Mechanics, 4, 25-38.
5. Suneja, B.P. and Datta, T.K. (1998) Active Control of ALP with Improved Performance Functions. Ocean Engineering, 25, 817-835.
6. Venkataramana, K., Kawano, K. and Yoshihara, S. (1998) Time-Domain Dynamic Analysis of Offshore Structures under Combined Wave and earthquake Loadings. Proceeding of the 8th International Offshore and Polar Engineering Conference, Montreal, 24-29 May 1998, 404-411.
7. Kawano, K. and Venkataramana, K. (1999) Dynamic Response and Reliability Analysis of Large Offshore Structures. Computer Methods in Applied Mechanics and Engineering, 168, 255-272.
8. Mahadik, A.S. and Jangid, R.S. (2003) Active Control of Offshore Jacket Platforms. International Shipbuilding Progress, 50, 277-295.
9. Abbasian-Hosseini, S.A, Hsiang, M.S, Leming, L.M. and Liu, M. (2014) From Social Network to Data Envelopment Analysis: Identifying Benchmarks at the Site Management Level. Journal of Construction Engineering and Management, 140, Article ID: 04014028.

10. Takewaki, I., Murakami, S., Fujita, K., Yoshitomi, S. and Tsuji, M. (2011) The 2011 off the Pacific Coast of Tohoku Earthquake and Response of High-Rise Buildings under Long Period Ground Motions. Soil Dynamics and Earthquake Engineering, 31, 1511-1528.

11. Abbasian-Hosseini, S.A., Liu, M. and Leming, L.M. (2015) Comparison of Least-Cost and Least-Pollution Equipment Fleet Configurations Using Computer Simulation. Journal of Management in Engineering, Article ID: 04015003.

12. Sorensen, R.M. (1997) Basic Coastal Engineering. 3rd Edition, Springer, New York.

13. American Petroleum Institute (2000) Recommended Practice for Planning, Design and Constructing Fixed Offshore Platforms—Working Stress Design. API Recommended Practice 2A-WSD, 21th Edition, USA.

14. Logan, D.L. (2011) A First Course in the Finite Element Method. 5th Edition, Cengage Learning, Essex County, New Jersey.

15. Fenton, J.D. (1985) A Fifth-Order Stokes Theory for Steady Waves. Journal Water Way, Port, Coastal & Ocean Engineering, 3, 216-234.

16. D'Alessandria, L.F. (1977) Design of the Self-Contained Fixed Offshore Oil Production Platform for the Shores of Venezuela. MSc. Thesis, MIT University, Cambridge, Massachusetts.

17. Abdel Raheem, S.E. (2013) Nonlinear Response of Fixed Jacket Offshore Platform under Structural and Wave Loads. Coupled Systems Mechanics, 2, 111-126.

3

CHAPTER

Design Considerations for Tension Leg Platform Wind Turbines

Erin E. Bachynski*, Torgeir Moan

Centre for Ships and Ocean Structures, NTNU, Marine Technology Centre, NO-7491 Trondheim, Norway

ABSTRACT

Tension leg platform wind turbines (TLPWTs) represent one potential method for accessing offshore wind resources in moderately deep water. Although numerous TLPWT designs have been studied and presented in the literature, there is little consensus regarding optimal design, and little information about the effect of various design variables on structural response. In this study, a wide range of parametric single-column TLPWT designs are analyzed in four different wind-wave conditions using the Simo, Riflex, and AeroDyn tools in a coupled analysis to evaluate platform motions and structural loads on the turbine components and tendons. The results indicate that there is a trade-off between performance in storm conditions, which improves with larger displacement, and cost, which increases approximately linearly with displacement. Motions perpendicular to the incoming wind and waves, especially in the parked configuration, may be critical for TLPWT designs with small displacement. Careful choice of natural period, diameter at the water line, ballast, pretension, and pontoon radius can be used to improve the TLPWT performance in different environmental conditions and water depths.

Keywords: Offshore wind energy; Tension leg platform; Parametric design; Dynamic analysis

1. INTRODUCTION

There is an increasing interest in using wind turbines offshore, and in deeper water. The tension leg platform wind turbine (TLPWT) concept is promising for intermediate water depths because the limited platform motions are expected to reduce the structural loading on the tower and blades compared to other floating concepts, without requiring the large draft of a spar or spread mooring system of a semi-submersible [1] and [2]. Although TLPWT concept design is an active research area [3], [4], [5] and [6], and some parametric studies based on simplified analysis are available, such as [7], [8] and [9], there is no clear consensus regarding optimal design, and there is great uncertainty regarding tendon design. TLPWT modeling must include sufficient detail to capture aerodynamic, hydrodynamic, and structural responses, but must also be able to efficiently examine a wide range of environmental conditions. Although oil and gas TLP design and analysis can provide some guidance, significant differences exist. For example, a major design difference between traditional TLPs and TLPWTs is that TLPWTs are not required to maintain a precise position relative to the seabed.

Existing independent 5 MW TLPWT designs include designs from Concept Marine Associates (CMA) [10], the Massachusetts Institute of Technology and the Italian Enel group (MIT/Enel) [11], MIT and the National Renewable Energy Laboratory (MIT/NREL) [3] and [12], the University of Maine (referred to here as UMaine)[2] and [13], Glosten Associates [5], GL Garrad Hassan (referred to here as GLGH) [4], a Japanese collaboration (Nihei et al.) [14], and IDEAS [15]. These designs represent a variety of displacements (Δ) from 846 tonnes [14] to 12,187 tonnes [3], with stiffness provided by 3–8 tendons, and diameter at the water line between 4.5 and 18 m. In addition to these designs, Crozier's master's thesis [9], which is focused on a 10 MW design, also includes an optimized, self-stable TLPWT design for the NREL 5 MW wind turbine with a displacement of 6360 tonnes. Interestingly, the self-stable 5 MW and 10 MW designs were very similar despite the increased turbine capacity [9]. All of the previously mentioned designs use vertical tendons, except the MIT/Enel tension leg buoy (TLBWT) [11], which uses tendons at approximately 45° inclination in order to limit horizontal plane motions as well as vertical motions; all are single-column designs, except for the CMA and IDEAS designs [15] that introduce several additional inclined columns that pierce the surface (to limit vortex-induced motions and reduce wave loads). The IDEAS design also introduces lateral taught moorings (to limit the mean offset and platform set-down). The pontoon volume is small (negligible) for the MIT/Enel, MIT/NREL, and Crozier designs, but is more significant for the other TLPWTs. In addition to these 5 MW designs, several multi-column designs exist for different turbine capacities. The EsaFloater (Δ = 3113 tonnes) described by Casale et al. uses a hexagonal TLPWT hull to support a 6 MW turbine [6], and Suzuki et al.'s 4100 tonne TLPWT for the 2.4 MW turbine has three additional surface-piercing columns for buoyancy, stability, and easy access to the tendon attachment points [16].

Although many of the published designs include descriptions of the hull shape, few details on tendon design are given. The tendons are assumed to be constructed as steel pipes in most of the above designs, except for the CMA design, which uses wire ropes [10], and the Glosten Associates design, which employs chain moorings [5]. Using this unique method, the cost of electricity was estimated to be comparable to sea-based bottom-fixed solutions for water depth greater than 50 m [5]. Note that the UMaine design, although inspired by [5], uses neutrally buoyant steel pipe tendons. GLGH also suggests wire rope and synthetic fiber lines as possible solutions [4]. For the foundation at the seabed, Casale et al. considered piles, suction buckets and gravity anchors, and concluded in favor of piles [6].

Computational analyses of different levels of complexity have been performed on all of the aforementioned designs, which have shown promising performance characteristics in most conditions. Robertson and Jonkman compared the UMaine and MIT/NREL concepts numerically using identical environmental conditions and determined that the responses, despite significant design differences, were quite similar as far as the structural loads on the wind turbine are concerned [2]. Limited small-scale test results have been published for the Japanese concept [14] and the UMaine TLPWT [13]. Compared to results using the FAST [17] computational tool with tuned damping coefficients, the experimental results in Ref. [13] show larger wave-frequency out-of-plane motions than predicted by FAST, and poor agreement in the high-frequency loading. A slack-line incident was observed both experimentally and numerically for an extreme wind and wave condition. In Nihei et al.'s experiments, large yaw, leading to capsize, was observed for the smaller of the two models [14]. Henderson et al. suggest that increased yaw stiffness may be sufficient to avoid this type of yaw instability, even when one blade experiences pitching mechanism failure [4].

Although frequency-domain methods are commonly used in preliminary design analysis for offshore platforms, coupled non-linear analysis in the time-domain is currently preferred for floating wind turbines due to the complex interactions between wind- and wave-driven responses. It is necessary to include sufficiently advanced structural, hydrodynamic, aerodynamic, and control models to capture these effects, while still managing computation time. Experience from the oil and gas industry indicates that slow-drift and sum-frequency forces may play a role in tendon loading [18], but TLP mooring loads are primarily linked to first order wave loads [19, chp 7.6.3]. In addition to the calculation of wind forces, analysis of TLPWTs differs from traditional TLP analysis due to the smaller size of the platform, shallower water depth, large pitch moment due to the wind turbine thrust force, the wind turbine loads due to aerodynamic and generator torque and rotational inertia, the blade pitch and generator torque control system, and the coupling effects between the structural, aerodynamic, and hydrodynamic responses. In an analysis of a small 1.5 MW TLPWT in 80 and 200 m water depths, Bae observed fewer coupling effects in shallower water and noted that second order effects on the tower could be critical [20]. Additionally, different combinations of environmental conditions must be considered. In addition to wind, wave, and current conditions, Suzuki et al.'s analysis even includes considerations for seismic loading that are particular to the proposed location [16].

The current work uses state-of-the-art computational TLPWT models to examine the effects of parameter variation on TLPWT performance. Five baseline parametric TLPWT designs were established using spreadsheet-based calculations, such that they met minimum design requirements. Based on these five designs, 40 additional designs were examined by varying the center column diameter, pontoon radius, permanent ballast weight, and water depth. The performance of the designs in operational and parked conditions with turbulent wind and irregular waves was compared by examining structural loads on the wind turbine and tendons, platform motions, and power production. The design and analysis bases are presented first, followed by an examination of the cost and performance of the baseline designs, and then an examination of the parameter sensitivity. Finally, overall trends are discussed in order to guide future TLPWT design and analysis.

2. PARAMETRIC TLPWT DEFINITION

As in Ref. [21], the design space considered here includes single-column TLPWTs with three or four "spokes" or pontoons, intended to support the NREL 5 MW wind turbine with the OC3 Hywind tower design [22] and [23]. The hull, made of steel, consists of a main cylinder (diameter D_1, length h_1) continued into a base node (diameter D_2, lengthh_2), with overall draft T. Protruding from one of these cylinders at a vertical location zs arenp rectangular or circular pontoons that support the tension legs, as shown in Fig. 1. For example, a TLPWT that resembles the Sea Star TLPs [24] with rectangular pontoons (height hp, width wp, and total radius from the cylinder center rp) and nt tendons is depicted in Fig. 1b. These pontoons are located at the base of the TLPWT and may provide substantial buoyancy. In contrast, Fig. 1c shows another alternative using thin struts with circular cross-section (diameter *dp* and total radius from the cylinder center *rp*) and *nt* tendons. In order to transfer the tendon forces to the hull, additional support wires may be required, but the structural details of these attachments are not considered in the initial design.

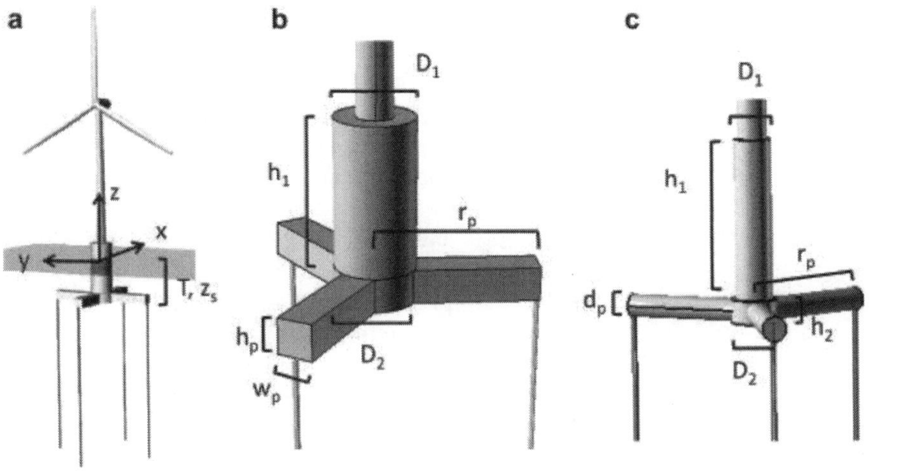

Figure 1. Parametric design definitions.

The tendons are assumed to be hollow circular pipes, characterized by their Young's modulus *Et*, cross-sectional area *At*, unstretched length l_0, dry mass per length *ρt*, and outer diameter *dt*. In general, it is desirable for the tendons to have near zero weight in water, which for typical steel density implies a thickness $tt \approx 0.033dt$.

In order to reduce the number of design parameters, the freeboard to the tower base ($bt = h_1 + h_2 - T$) is constant for all designs (10 m). Some permanent concrete ballast may be added to the base node and center column in order to improve the performance or allow greater stability during installation. The parameter BF is defined as the ballast mass divided by the displaced water mass of the center columns and no ballast is considered in the pontoons.

The coordinate system origin is at the center of the single column at the still water level, with *z* vertically upward and *x* in the downwind direction, which is also the wave propagation direction in this study. The first pontoon or strut is oriented in the *x* direction.

2.1. Spreadsheet design calculations

Based on the previously described parameters, first predictions of TLPWT behavior can be obtained through simple spreadsheet computations, following engineering assumptions and physical principles. These spreadsheet calculations were used to identify candidate designs that meet the criteria described in Section 2.2.

The displacement was easily found from the geometry, while approximate values for the steel weight in the hull are found based on assumed steel weights per volume Table 1[25]. By further assuming that the steel weight is distributed as cylinders and boxes of uniform thickness, the 6 × 6 mass and inertia matrix [*M*] was filled. Simple linear approximations were then used to find the added mass matrix [*A*], hydrostatic stiffness matrix [*C*], and mooring system stiffness matrix [*K*] based on the geometry and pretension, following for example [26], [27] and [28]. The added mass is estimated by summing the 2-D cross-sectional properties of the main columns and pontoons, without considering any interaction. The hydrostatic stiffness in heave is a function of the waterplane area, while the pitch and roll hydrostatic stiffness depends on the waterplane moment of inertia, center of buoyancy, and center of gravity. The mooring system stiffness can be approximated by assuming that the tendons remain straight and disregarding changes in buoyancy. Additional details may be found in the Appendix.

Table 1. Weight approximations for structural steel [25].

Column 1 (upper)	157 kg/m^3
Column 2 (lower)	224 kg/m^3
Pontoons	202 kg/m^3

2.2. Design criteria

The overall goal of TLPWT hull design should be to minimize the cost of electricity: that is, to maximize power production and minimize construction, installation and operational costs. A complete calculation of the cost of electricity is not realistic, but one can consider several design and performance characteristics that may affect cost. In order to minimize the construction and material cost, the steel mass, tendon pretension, and displacement should be minimized; to limit the operational costs, the tendon, tower, blade, and nacelle loads and load variations should be minimized. The installation costs, which depend heavily on vessel rent and seabed analysis, are assumed to be similar for all designs, and are neither estimated nor compared here.

With these goals in mind, the following criteria were applied at the spreadsheet design stage:

1.The surge and sway natural periods should be longer than 25 s in order to avoid first order wave excitation. The decoupled undamped surge natural period Tn_1 may be estimated as in Eq. (1).

$$T_{n1} = 2\pi\sqrt{\frac{M_{11} + A_{11}}{K_{11}}} \tag{1}$$

2.The heave, roll/bending, and pitch/bending natural periods should be shorter than 3.5 s in order to avoid first order wave excitation. The rigid body natural periods are approximated by, for example, Eq. (2), including the hydrostatic stiffness terms. The effect of tower bending is not including in the spreadsheet design stage, but is checked in the finite element model used in the time-domain analysis.

$$T_{n3} = 2\pi\sqrt{\frac{M_{33} + A_{33}}{C_{33} + K_{33}}} \tag{2}$$

3.As in a typical TLP design process, in order to limit the angle at the tendon connectors, the mean offset should not exceed 5% of the water depth [19, chp 7.6.3]. The mean offset is a function of the mooring system stiffness and the mean load due to wind, waves, and current. In the preliminary design, the wind effect is taken to be the maximum steady turbine thrust, the mean surge force due to waves is estimated from the pressure integration and viscous forces [28] as in Eq. (3) for regular waves with the significant amplitude ζa, and current is neglected. The mean offset is estimated for significant wave height 4 m and period 10 s, which corresponds approximately to the wind speed that generates the maximum thrust, using drag coefficient CD = 1.

$$F_1 = \frac{2}{3}\rho g \zeta_a^2 \frac{D_1}{2} + \frac{2}{3\pi}\rho C_D D_1 \omega^2 \zeta_a^3 \tag{3}$$

4.The tendon area must be sufficient to prevent yield (σy), within a given safety factor (SF), for tensions up to twice the initial tension (Tt), as in Eq. (4). In this study, SF = 2. This leads to a conservative design, accounting for one tendon being very near slack, and with a high safety factor compared to, for example, the API Recommended Practice 2T [29].

$$\frac{2T_t}{A_t} \leq \frac{\sigma_y}{\mathrm{SF}} \tag{4}$$

5.$\Delta \geq 2000$ m^3, based on previous results that indicate that a minimum displacement may be required to survive extreme wind and wave conditions [21]. A minimum displacement increases both the stiffness of the system, due to higher pretension, and its ability to mitigate extreme loads by platform inertia.

Inherent stability during the float-out phase is not a design requirement in the current study, so the installation will require additional consideration. The important coupled pitch/tower bending and roll/tower bending modes of the TLPWT system cannot be predicted by the spreadsheet calculations, but further constraints on the natural periods should consider 1p and 3p blade passing excitation. The maximum frequency for 1p excitation is 0.2 Hz, while the minimum frequency for the 3p excitation is approximately 0.37 Hz. In order to avoid both wave and 3p excitation, the coupled pitch/bending mode should then ideally fall between 2.7 s and 3.5 s.

3. NONLINEAR COUPLED ANALYSIS

Three computer codes are used to model the coupled behavior of the TLPWT systems in the time domain: Simo, which models the rigid body hydrodynamics of the hull [30]; Riflex, which includes the finite element solver, flexible elements for the tendons, tower, shaft, and blades, and the link to an external controller [31]; and AeroDyn, which provides the forces and moments on the blades based on Blade Element/Momentum (BEM) or Generalized Dynamic Wake (GDW) theories, including dynamic stall, tower shadow, and skewed inflow correction [32]. The generator torque and blade pitch control system is written in Java. This combination provides a stable nonlinear finite element solver, sophisticated hydrodynamics, well-tested aerodynamics, and control logic. The Simo–Riflex wind turbine module has been previously verified [33] and [34], and the Simo–Riflex-AeroDyn combination is documented in Ref. [35].

3.1. Environmental conditions

Six environmental conditions (ECs) with directionally aligned turbulent wind and irregular waves were selected as test cases for the TLPWT performance, as described in Table 2. The first four conditions represent simultaneous increases in wind speed and wave severity, and follow from the joint distribution described in Ref. [36]. EC 1 represents a below-rated wind speed case with high turbulence and relatively small waves, while the average wind speed in EC 2 is the rated speed (11.4 m/s). ECs 3 and 4 represent an above-rated condition where the generator is operational, and an extreme storm condition where the turbine is parked with feathered blades, respectively. The two final load conditions are representative of DLC1.6 and DLC1.3 in Ref. [37], respectively. EC 5 considers an operational condition in storm waves, while EC 6 considers extreme turbulence near the wind speed that generates maximum thrust.

Table 2. Environmental conditions.

Condition	1	2	3	4	5	6
Significant wave height, H_s (m)	2.5	3.1	4.4	12.7	12.7	3.1
Peak wave period, T_p (s)	9.8	10.1	10.6	14.1	14.1	10.1
Mean wind speed at hub, U (m/s)	8.0	11.4	18.0	50.0	18.0	12.0
Turbulence intensity at hub, I	0.20	0.17	0.15	0.11	0.15	0.26

Six wind and wave time series of 2000 s duration, without repetition, were generated for each of the six environmental conditions using different seeds. These six conditions were then applied to each of the TLPWT designs. Removing the initial transients from the simulations, a total of 3 h of simulation time with identical environmental conditions were used to compare the designs. A comparison of baseline designs using six 3-h simulations indicated that statistics from a single 3-h simulation were sufficient to compare the parametric designs.

The simulated wave series were generated based on JONSWAP spectra, where the peak enhancement factor depends on the significant wave height (Hs) and peak period (Tp) [30]. A frequency discretization of 0.002 rad/s was used.

The three-dimensional wind fields were generated in NREL's TurbSim [38] based on the Kaimal spectrum with the IEC Class B normal turbulence model (NTM) [39]. The wind shear was modeled by the power law with exponent 0.14 and surface roughness length 0.3 mm [37]. In the vertical plane 32 × 32 points were used, with wind field generation time step 0.05 s.

3.2. Hydrodynamic modeling

In the simulations presented here, the potential flow model includes only the first order wave forces and the mean drift forces based on the first order solution. In order to obtain the potential flow theory added mass, damping, and excitation forces, a first order panel model was generated for each design in Wadam (Det Norske Veritas 2010). The full hull was included in the panel model except for TLPWT 1, where the pontoons were only modeled by Morison elements (including added mass). For all TLPWT designs, additional Morison-type viscous forces were applied to the main column and spokes, using quadratic drag with $CD = 0.7$ for tangential flow and $CD = 0.07$ in the axial direction[28]. For the rectangular spokes, the quadratic drag was applied as though the section were circular with diameter equal to the average horizontal dimension ($[hp + wp]/2$). This may underpredict the viscous drag on the rectangular sections, but gives a reasonable approximation.

3.3. Structural modeling

The columns and pontoons were considered as rigid bodies. The tower, shaft, and tendons were modeled with flexible axisymmetric beam elements, while the blades were modeled with flexible beam elements with two symmetry planes to capture the difference in stiffness in the edgewise and flapwise directions. The blade model accounts for geometric stiffening, but assumes that the center of gravity, shear center, and elastic center are coincident. Despite these simplifications, code-to-code comparisons of the aeroelastic response of the NREL 5 MW turbine have shown very good agreement between state-of-the-art codes, including Simo–Riflex–AeroDyn [35]. The boundary conditions for the tendons were a simple support at the anchor and a more complex joint at the fairlead connection to the rigid hull: fixed in relative translation and tendon torsional rotation, but free in tendon bending rotation.

The equations of motion were solved by Newmark-beta numerical integration ($\beta = 0.256, \gamma = 0.505$) with a time step of 0.005 s. Structural damping, which is needed for numerical stability and can significantly affect the results, was included through stiffness-proportional Rayleigh damping terms. The tower and tendon damping coefficient was set to 0.001s multiplied by the local stiffness (K), while the blade damping coefficient was

0.002s*K, and other components had a damping coefficient of 0.005s*K. These coefficients gave good numerical stability and provided a reasonable approximation to the damping coefficients specified in Ref. [22] for the first two isolated blade modes and the first two isolated tower modes. Furthermore, using the chosen coefficients, the tendon structural damping was below 1% for modes as short as 0.31s, which was shorter than all of the rigid body modes and the first two coupled pitch/tower bending modes, and reasonable for large steel structures [40].

3.4. Aerodynamic modeling

The AeroDyn code provides both BEM and GDW options, and an attempt was made to use appropriate models for each wind speed and operating conditions. For mean wind speeds 8 m/s or 50 m/s, the BEM model was used, but the induced velocity iteration was bypassed for the parked turbine at 50 m/s. For wind speeds 11.4 m/s or 18 m/s, the GDW model was applied to capture the dynamic wake effects. The Beddoes-Leishmann dynamic stall model was applied to all simulations.

3.5. Control system modeling

The prescribed NREL baseline controller is described in Ref. [22]. The generator speed filter and generator torque lookup table were fully implemented, and the PI control for blade pitch was included with the same frequencies as in the land-based controller. Floating wind turbines may experience a negative feedback loop related to nacelle velocity above rated speed [41] and [42]. Although the pitch natural frequencies are high enough to largely avoid this instability in pitch, TLPWTs may experience instability due to surge motions [3]. Because the surge damping was sufficiently large, the negative feedback was not found to be a problem.

4. BASELINE DESIGNS

Table 3 and Fig. 2 present the five baseline designs used in the parametric study, in order of decreasing displacement. These designs were not optimized, but were established based on the spreadsheet design procedure, and all met the design criteria described in Section 2.2. The parametric variations of these designs also met the design criteria, except for TLPWT 5 with decreased pontoon length (Δ too low at 1937 m^3).

Table 3. Baseline designs (water depth 150 m).

	TLPWT 1	TLPWT 2	TLPWT 3	TLPWT 4	TLPWT 5
Pontoons	Rect.	Rect.	Rect.	Rect.	Round
D_1 (m)	18.0	14.0	14.0	6.5	6.5
D_2 (m)	18.0	14.0	14.0	10.0	6.5
h_1 (m)	52.6	40.0	26.0	33.0	23.0
h_2 (m)	2.4	5.0	6.0	6.0	5.0
r_p (m)	27.0	32.0	28.0	25.0	32.5
h_p/w_p or d_p (m)	2.4/2.4	5.0/5.0	6.0/6.0	6.0/6.0	5.0
BF	0.55	0.60	0.40	0.40	0.40
n_p	4	3	3	4	3
z_s	−43.8	−32.5	−19.0	−19.0	−15.5
d_t (m)	1.4	1.1	1.3	1.2	0.9
t_t (mm)	46.2	36.3	42.9	39.6	29.7
Steel mass (tonnes)	2322	1518	1293	859	505
V (m^3)	11,866	7263	5655	4114	2320
T_t (kN)	6868	4963	8262	5556	3384

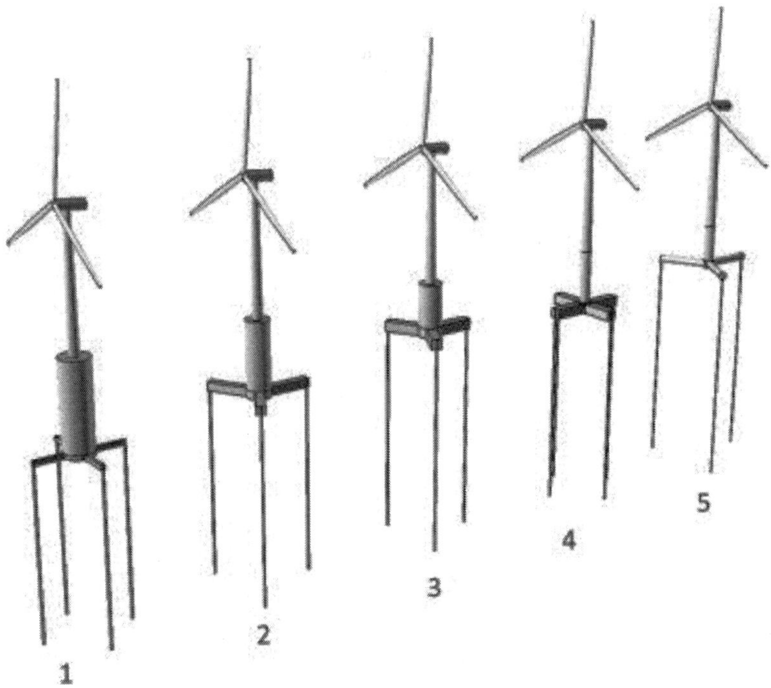

Figure 2. Baseline TLPWT designs.

The first design (TLPWT 1) was based on the NREL-MIT design described in Denis Mathas master's thesis [3]. The mass matrix was significantly modified, however, due to the different steel mass estimation method, and the tendon arrangement was simplified such that only one tendon was considered on each pontoon. In order to satisfy design criteria 3 and 4, the tendon stiffness was increased compared to the original design. TLPWT 1, which had the largest displacement by far, depended almost exclusively on the center column for buoyancy. The first design was the only design that would be stable in tow-out, including the wind turbine and appropriate water ballast in the pontoons and center column to achieve the design draft.

TLPWT 2 was a 3-legged platform with approximately 60% of the displacement and 70% of the initial tendon tension of the first design. Approximately 30% of TLPWT 2's displacement came from the pontoons, making it unstable for tow-out.

The hull of TLPWT 3 was an approximately half-scale version of the original Sea Star oil platform [24]. Somewhat higher initial tendon tension was required to satisfy the third criterion. Approximately 60% of the displacement came from the center column, with the rest coming from the pontoons. Although TLPWT 3 had approximately half of the displacement of TLPWT 1, the pretension per line was significantly higher.

The fourth and fifth designs were inspired by the GLGH design [4], with some adaptations to support the OC3 wind turbine and satisfy the design criteria. The pontoons provided approximately 70% of the displacement, while the upper center column diameter was equal to the tower base diameter. In the parametric study, the diameter of the upper column was allowed to be less than the tower base diameter, although this is not realistic. The small column diameter was intended to make the structures more transparent to waves, but the low displacement also made TLPWTs 4 and 5 more sensitive to extreme conditions. TLPWT 4 had approximately twice the displacement and total pretension of TLPWT 5, but employed 4 rectangular pontoons instead of 3 round ones. The UMaine TLPWT design falls in between these two designs in terms of displacement and initial tendon tension, but has more flexible tendons.

The natural periods (*Tnj*) of the baseline designs for the rigid body motions in surge, heave, and yaw, and the two first coupled pitch-tower bending modes, obtained by decay tests using the full finite element model, are summarized in Table 4. TLPWTs 3 and 5 coincidentally showed very similar natural periods despite large differences in displacement and pretension. The coupled pitch-bending natural periods were very similar for all of the designs: due to the high stiffness in platform pitch (rigid modes <1.5 s), the tower-nacelle-rotor modes dominated. This study considered the hull to be a rigid structure, but the flexibility of the hull (particularly the pontoons) may change the natural frequency in heave as well as pitch [4]. Furthermore, the yaw natural frequency, particularly for smaller hull shapes, was found to depend significantly on the mass and added mass of the tendons.

5. RESULTS

5.1. Simplified cost comparison

A simplified cost comparison for the TLPWT hulls was performed following Crozier's model, as in Table 5[9]. This simple model considers upper and lower multiplicative factors for the steel mass, concrete ballast mass, and tendon pretension (indicative of anchor system cost). The resulting values are crude estimates for comparing the different designs, without considering the tower, rotor, and drive train costs. The cost model results for the baseline designs and several of the published designs are shown in Fig. 3.

Table 4. Baseline design damped natural periods, including tower flexibility, based on decay tests in the FE model (water depth 150 m). Land-based values for the 87.6 m tower are given for comparison (note that the TLPWT tower is 77.6 m).

	Land-based	TLPWT 1	TLPWT 2	TLPWT 3	TLPWT 4	TLPWT 5
Surge (s)	-	56.27	52.26	41.86	34.22	40.13
Heave (s)	-	0.55	0.75	0.60	0.52	0.60
Pitch/Bending (1) (s)	3.12	2.77	2.81	2.75	2.74	2.69
Pitch/Bending (2) (s)	0.35	0.51	0.48	0.39	0.39	0.39
Yaw (s)	-	13.99	17.97	18.63	19.83	19.59

Table 5. Cost estimation model, including lower and upper estimates.

	Lower	Upper	
Steel mass (hull and tendons)	600	800	$/tonne
Concrete ballast mass	50	150	$/tonne
Tendon pretension	15	25	$/kN

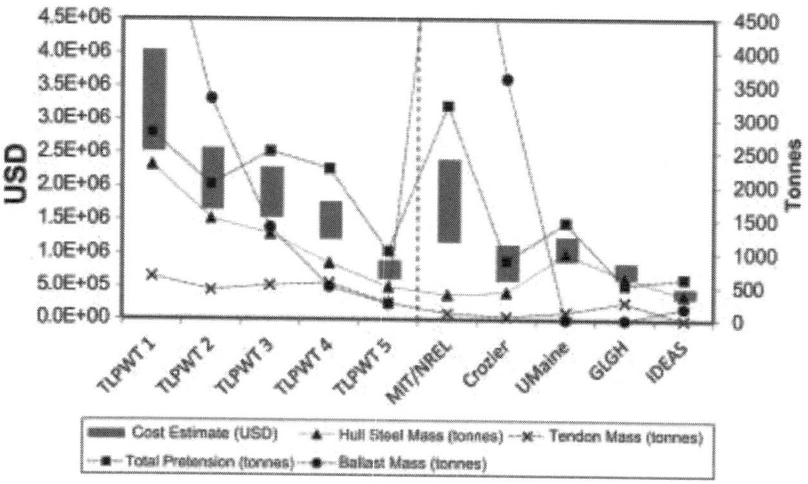

Figure 3. Hull and tendon simplified cost estimates, including weight/pretension breakdown, for the baseline designs and selected published designs. The ballast mass is 6456 tonnes for TLPWT 1 and 8220 tonnes for MIT/NREL.

The construction cost estimates for TLPWTs 1–5 (and parametric variations) were approximately linearly dependent on the displacement. The recomputed costs of the published platform designs also depended on displacement, but were found to be more variable due to differences in assumptions about hull scantlings and tendon weight per length. For example, the MIT/NREL hull design assumes an equivalent hull thickness of only 0.015 m, while the mass model in Table 1 would give an average hull thickness of more than twice that. Thus, despite the fact that the displacements of TLPWT 1 and MIT/NREL are similar, the cost estimates differed significantly due to the large differences in steel mass. Installation costs were not compared, but TLPWT 1 was the only self-stable construction out of the new TLPWT designs, and can therefore be expected to be somewhat easier to install than TLPWTs 2–5. Neither installation nor maintenance costs were estimated.

5.2. Performance comparison parameters

Several representative parameters – platform motions ($\zeta j, j = 1–6$), fore-aft tower base bending moment (M_{FA}), side–side tower base bending moment (M_{SS}), and tether tension in line i (Ti) – were taken as primary performance measures. As this study considered only aligned wind and wave loads, the in-plane motions (i.e., surge, heave, and pitch) and fore-aft tower base bending moment were expected to be the most important. In turbulent wind conditions, however, both roll and yaw motion may be excited by time–varying torque and asymmetric wind loads. These out-of-plane motions were coupled via the thrust force and were not reduced significantly by aerodynamic damping. The power production (P) of the different designs was also examined, but the results indicated very little difference among the different designs in that respect. The standard deviation (σ) in the different environmental conditions was used as the main comparison measure, rather than predicting extreme values or fatigue cycles.

5.3. Comparison of baseline TLPWT designs

First, the mean values (μ) of different performance indicators are shown in Fig. 4. The mean values of surge and pitch, $\mu(\zeta_1)$ and $\mu(\zeta_5)$, depended on the mooring system stiffness as well as the wind and wave forcing. The surge stiffness of TLPWT 5 was less than half that of TLPWTs 2–4 and approximately one-third of TLPWT 1. Because the mean wave force was small compared to the thrust force (or drag force for the parked case), $\mu(\zeta_1)$ was approximately inversely dependent on the stiffness, and varied more with wind speed than wave height. Similarly, $\mu(\zeta_5)$ was primarily dependent on the mooring system pitch stiffness and the moment due to the wind turbine thrust or drag. TLPWT 1 had the largest K_{55}, which was approximately 2.8 times the K_{55} of TLPWT 5, which had the softest mooring system. Because $\mu(\zeta_5)$ was small, $\mu(M_{FA})$ was not strongly affected by gravitational loading. As shown in Fig. 4, $\mu(M_{FA})$ did not vary significantly among the designs.

The mean tendon tension ($\mu(T_1)$, Fig. 4) was not necessarily equal to the still water value, because $\mu(\zeta_1)$ increased the tendon tension in all lines, while $\mu(\zeta_5)$ decreased the tension in the downwind line and increased the tension upwind. For line 1 (downwind),$\mu(T_1)$ can be approximated as:

$$\mu(T_1) \approx \frac{T_t}{\cos\theta_1} + \frac{\rho g D_1^2 \pi l_0 (1 - \cos\theta_1)}{4 n_t} - \frac{E_t A_t r_p \sin(\mu(\zeta_5))}{l_0} \quad (5)$$

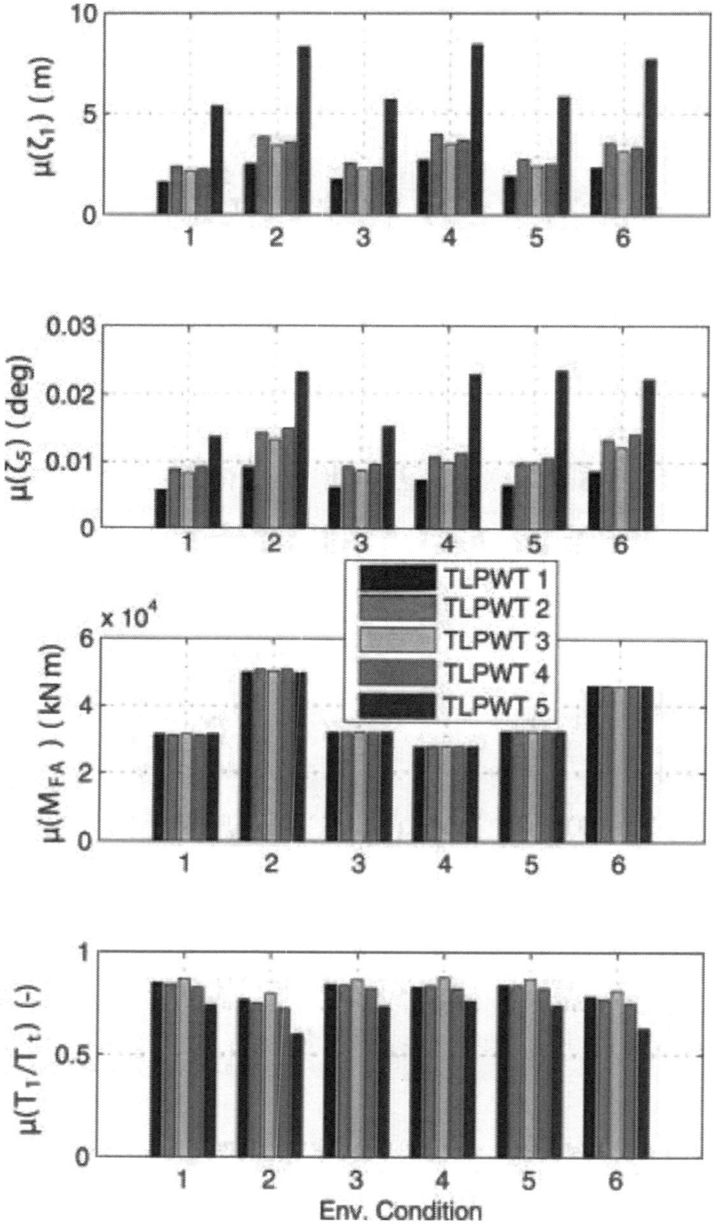

Figure 4. Comparison of baseline TLPWT mean surge, pitch, tower base bending moment, and line tension.

where $\theta_1 = \arcsin(\mu(\zeta_1)/l_0)$. The decrease in tension due to pitch was generally larger than the increase due to surge motion. The mean tendon tension is an important consideration when evaluating the probability of tendon slack events.

Fig. 5 provides an overview of the standard deviation of key performance indicators for the baseline designs in the four environmental conditions. While $\sigma(\zeta_1)$ was very similar for TLPWTs 1–4, it was somewhat higher for TLPWT 5 (particularly around the rated wind speed in ECs 2 and 6, where the thrust force is at a maximum). The low-frequency surge response was primarily excited by the wind, which was identical for all of the designs, and mitigated by platform inertia. While the surge response is not expected to be critical for the wind turbine performance or fatigue, large surge displacements may be critical for the tendon seabed connectors and the power cable due to large induced angles, and can also induce large tension variations due to set-down effects. Assuming that the tendon connectors are similar to those used in the offshore industry, for 150 m water depth, the maximum displacement should not exceed 15 m [19, chp 7.6.3].

Although $\sigma(\zeta_5)$ was quite small for all of the designs, TLPWT 5 showed significantly larger low-frequency (wind-induced) and wave-frequency response, the latter of which became increasingly important as the wave height increased. Comparing the pitch standard deviation in ECs 4 and 5, the operational turbine is seen to increase the pitch motions compared to the parked case.

M_{FA} depended on the nacelle motions (gravitational loading) relative to the platform, the thrust force, and the load transfer to the hull inertia and tendon tension. As shown inFig. 5, the platform pitch motions were not a good predictor of M_{FA}. Significant contributions to $\sigma(M_{FA})$ occurred in three frequency ranges: low-frequency (wind-induced), wave-frequency (wave-induced), and at the first coupled pitch-bending frequency (structural). As an example, the spectrum of M_{FA} for EC 2 is shown in Fig. 6. Wave-frequency loading dominated M_{FA}, indicating that gravity loads were important even though the TLPWT motions were small. The contributions to the bending moment at the first pitch/bending natural period and at 3p may have an impact on fatigue. The wave-frequency tower bending moment, which varied most among the different designs, was lowest for TLPWTs 1, 2, and 4, while the loading near the natural period was more similar. Furthermore, the wave-frequency tower bending moment of TLPWT 3 was larger than expected based on the general trend toward smaller tower loading as the displacement increases. This may be related to the frequency dependence of the wave excitation of platform pitch: the wave forcing transfer function for TLPWT 3 had its peak at a higher frequency. EC 5 was also seen to be a critical load condition for M_{FA}, where the operating turbine increased the loading compared to EC 4.

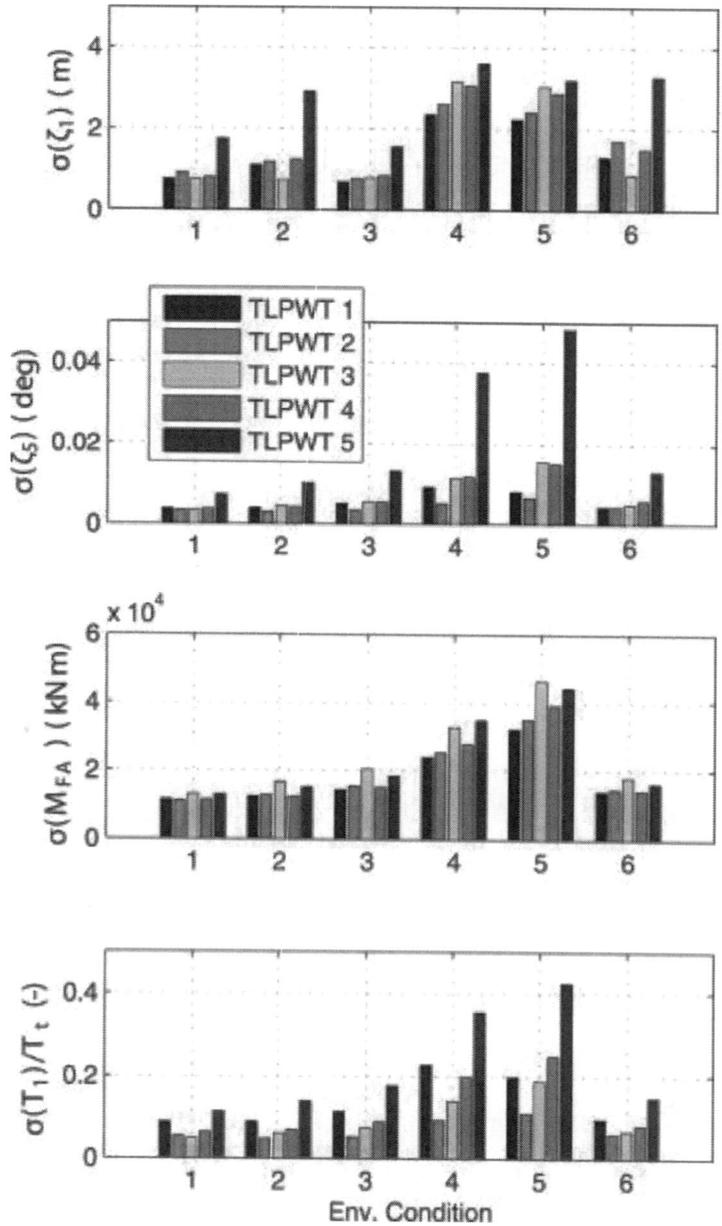

Figure 5. Comparison of baseline TLPWT standard deviations for surge, pitch, tower base bending moment, and line tension.

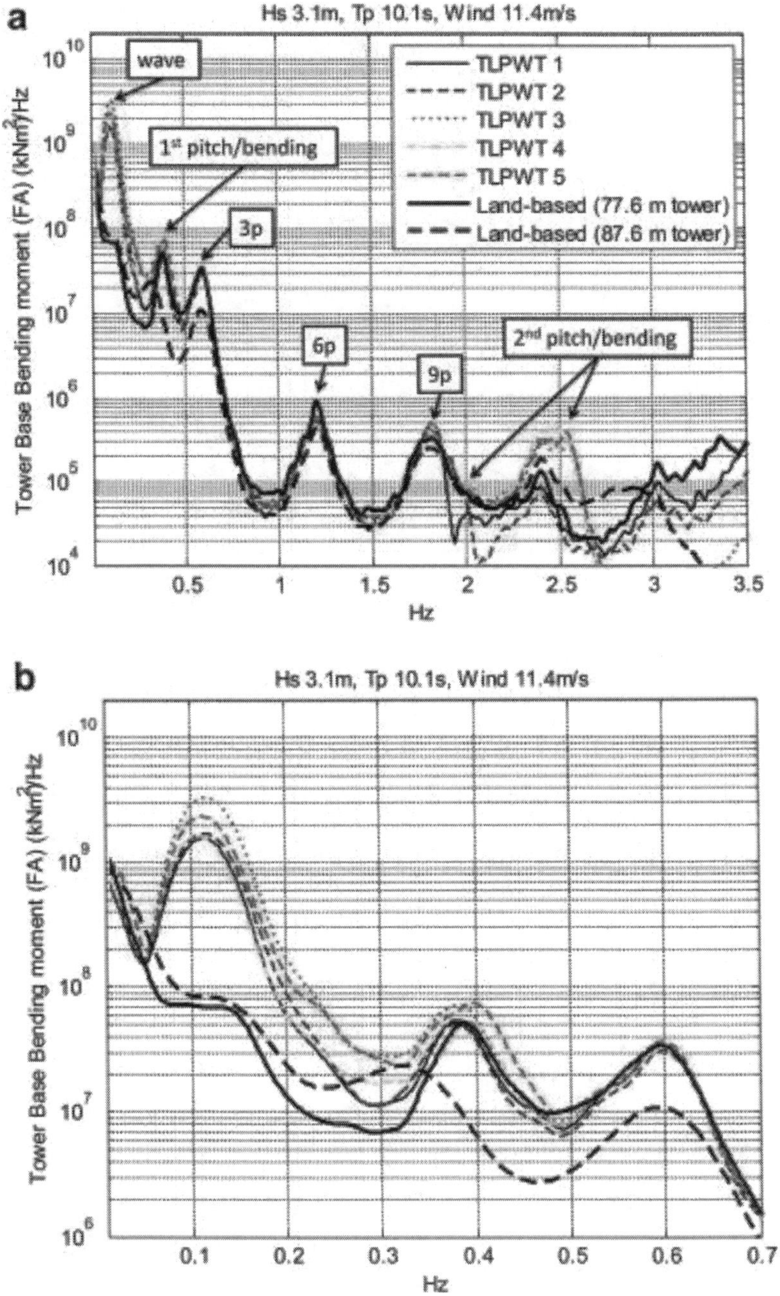

Figure 6. Spectrum of M_{FA} for EC 2, including zoomed view around the wave frequency (b). Note that M_{FA} is measured 10 m above still water/ground level for all cases.

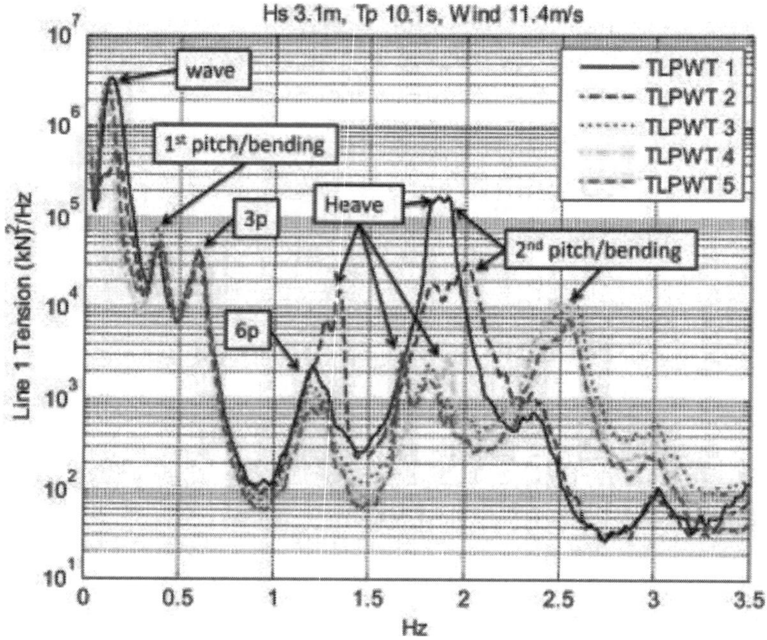

Figure 7. Spectrum of T_1 for EC 2.

The standard deviation of line 1 tension, $\sigma(T_1)$, is an important consideration when evaluating the probability of tendon slack and the fatigue life of the tendons. Fig. 5 shows the standard deviation of the line tension as a fraction of the line pretension. Assuming tendon tension to be a Gaussian process, in order to avoid slack conditions, we require $([\mu(T_1) - k\sigma(T_1)]/Tt) > 0$, where the factor k should be chosen to represent an acceptable probability. Using the limited conditions considered here, it is not possible to give a single value of k that reflects the full range of environmental conditions that may be encountered during the TLPWT lifetime, TLPWTs 1–4 showed k values of at least 3.3 in all of the considered cases. Assuming a Gaussian distribution, the probability of slack in environmental condition 5 for TLPWT 5 is approximately 4.25%, indicating that the design may require further modification. The variation in T_1 did not appear to be very sensitive to the wind speed when the turbine was operating, but increased with the wave height in the extreme environmental conditions, EC 4 and 5.

Fig. 7 shows the spectral response of T_1 in EC 3, and highlights the components at the wave, structural, and wind turbine excitation frequencies. As shown, the heave and second pitch/bending natural frequencies of TLPWT 1 were very similar, which contributed to a larger peak in the spectrum of T_1. The wave-frequency tension variation for

TLPWT 2 was significantly smaller than that of the other designs. As shown, the relative importance of the wind turbine rotation frequencies depended on the design.

5.4. Comparison of baseline designs with the land-based turbine

Fig. 8 compares the means and standard deviations in M_{FA}, M_{Oop}, M_{Ip}, and P for the baseline TLPWT designs as a function of the land-based (77.6 m tower) turbine in the same wind conditions. Note that the original land-based design is supported by an 87.6 m tower [22], but the OC3 Hywind tower [23] was modeled as though there were rigid foundation at 10 m height for this comparison. The results for the land-based turbine are summarized in Table 6.

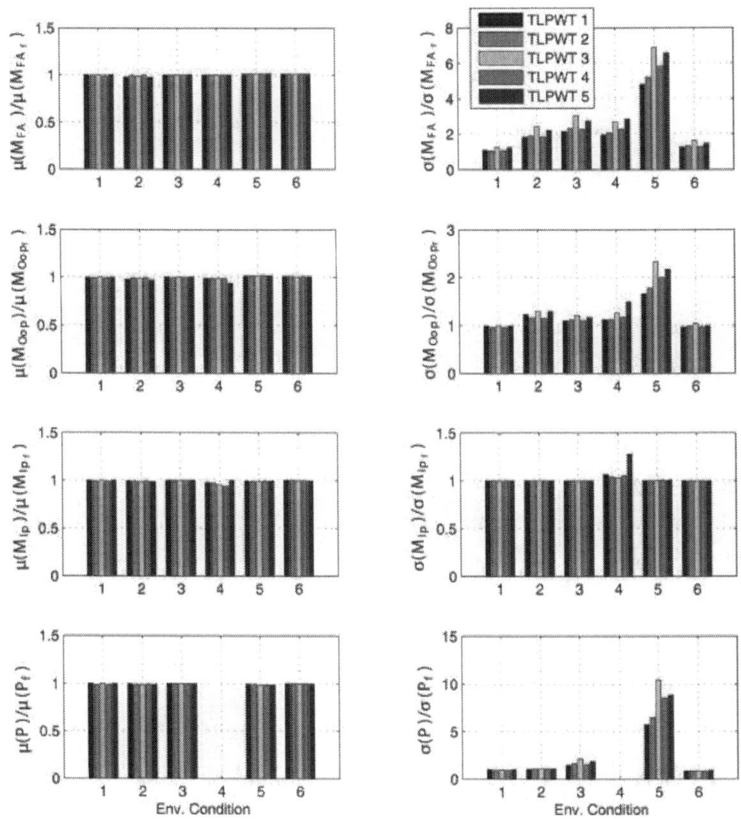

Figure 8. Comparison of baseline TLPWT and fixed turbine (77.6 m tower) loads and power generation.

Table 6. Fixed turbine with 77.6 m tower: mean and standard deviation of loads and power generation. Note that EC 5 results are identical to EC 3.

	EC 1		EC 2		EC 3		EC 4		EC 6	
	μ	σ	μ	σ	μ	σ	μ	σ	μ	σ
M_{FA} (kN m)	31,299	10,744	50,910	6888	31,970	6708	27,741	12,401	45,517	11,240
M_{Oop} (kN m)	5183	1562	8719	1350	4517	1682	335	424	7740	2190
M_{Ip} (kN m)	634	2572	1265	2586	1316	2628	−1339	1157	1253	2609
P (kW)	1911	800	4714	487	5000	33	0	0	4670	698

As shown in Fig. 8, the mean bending moments at the tower base and at the blade root were nearly identical to those seen in the land-based turbine. For EC 4, the blade loads are given for the blade that is fixed in a vertically upward position. The mean power, $\mu(P)$, was also well-regulated by the controller and was not affected by the TLPWT motions in waves.

The structural loading and power production were, however, more variable for the TLPWTs than the land-based structure. The effect on the blade root moments was relatively small, but $\sigma(M_{FA})$ was significantly higher than in the land-based case, especially as the wind speed increased, and particularly in EC 5 where storm waves were encountered by an operating turbine. It is important to note that the increase in$\sigma(M_{FA})$ occurred at the wave-frequency, which has a lower encounter frequency than the primary bending moment variation for the land-based wind turbine. Previous studies of equivalent loading indicate increased design equivalent loading on the order of 1.2 times that of the land-based tower [2]. Furthermore, the power variation appears to increase significantly in EC 3 and 5, but it is important to note that $\sigma(Pf)$ is very close to zero, making a small increase in the absolute value appear as a large relative change.

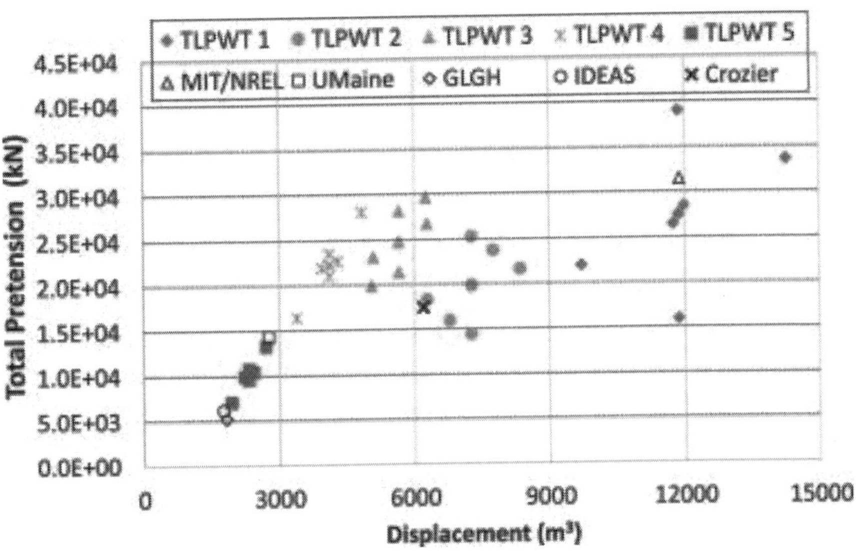

Figure 9. Total pretension vs. displacement for all of the TLPWT design variations, including comparisons with published designs [3], [2], [4], [15] and [9].

5.5. Parameter variations

Variations in the TLPWT hull diameters, water depth, pontoon radius, and ballast were considered for all of the baseline designs. The range of displacements and total pretension represented by the different design variations is summarized in Fig. 9. As shown, the designs considered here are comparable to several published designs.

The trends in behavior observed for the variations are summarized in Table 7 and in the sections below.

Table 7. Effects of increases in D_1 and D_2, H, rp, and BF on performance parameters. + indicates improved performance, – indicates decreased performance, and 0 indicates no clear trend, with the number of + or – symbols indicating stronger effects. Where the effect of increasing D_1 and D_2, H, rp, or BF is dependent on the baseline design displacement (Δ) or wave condition (*Hs*), symbols are given in order of increasing Δ or *Hs*.

	D_1 and D_2	H	r_p	BF
$\sigma(M_{FA})$	− −	+ +	0	+ +
$\sigma(T_1/T)$	+	+/− (Δ)	+ + +	− −
$\sigma(M_{Oop})$	−	+ +	0	+ +
$\sigma(\zeta_1)$	0	+/− (Δ)	+/− (H_s)	+/− (H_s)
$\sigma(\zeta_5)$	+/− (Δ)	− − −	+ + +	+/− (Δ)
$\sigma(\zeta_4)$	+ +	− −	+ + +	− − −
$\sigma(\zeta_6)$	+ +	− −	+ + +	− − −

5.5.1. Diameter

The first two variations applied to the TLPWT designs were a 10% increase and a 10% decrease in the main column diameters, D_1 and D_2. An increase in the diameter by a factor λ increased the column displacement and mass by λ^2. A_{11}, A_{44}, and A_{55} also increased by λ^2, while A_{33} increased by λ^3. The resulting increase in inertia and pretension resulted in little net effect on the surge and yaw natural periods. For the TLPWTs that did not depend heavily on the main column for displacement, changing the diameter by 10% had little effect on the natural periods. On the other hand, for a smaller platform that depended primarily on the column for displacement, the increase in pretension due to an increase in the D_1 and D_2 could be significant, particularly if the unchanging turbine mass represented a significant component of the overall mass.

The effect of diameter changes on the surge motion depended heavily on the baseline concept and the environmental conditions, and did not show any clear trends. For example, $\sigma(\zeta_1)$ increased 10–20% for TLPWTs 2 and 3 for both a 10% increase and a 10% decrease in D_1 and D_2. The increased diameter resulted in longer surge natural periods, which were primarily excited by the wind forces, while the decreased diameter resulted in decreased pretension, and lower stiffness, which gave larger motion responses.

The effect on the pitch motions was also complicated. The increase in D_1 and D_2 resulted in 10–20% increases in $\sigma(\zeta_5)$ for TLPWTs 1 and 2, and less than 10% decreases in $\sigma(\zeta_1)$ for the other designs. A corresponding decrease in D_1 and D_2 reduced the pitch motions of TLPWT 1, but slightly increased the pitch motions of the other designs. The effects on the out-of-plane motions were more consistent: an increase in D_1 and D_2 resulted in decreases in both $\sigma(\zeta_4)$ and $\sigma(\zeta_6)$.

The structural loading also showed varying degrees of sensitivity to D_1 and D_2. Although a 10% decrease in D_1 and D_2 resulted in an increase in $\sigma(T_1)$ of up to 10% for TLPWTs 2 and 3, and smaller increases for the others, a corresponding increase in D_1 and D_2 had more mixed impacts on $\sigma(T_1)$. The increase in diameter decreased $\sigma(T_1)$ by 5–10% for all of the designs except TLPWT 2, which showed up to 10% more surge motion, particularly in EC 4. The variation in D_1 and D_2 resulted in less than 4% change in $\sigma(M_{FA})$, with an overall increase in $\sigma(M_{FA})$ for increased D_1 and D_2.

5.5.2. Water depth

TLPWT platforms are most likely to be considered for water depths between 70 m (an approximate upper limit for bottom-fixed platforms) and 200 m (where spar platforms are likely to have economic advantages). The performance of the baseline designs was examined for 100 m and 200 m water depths, in addition to the design depth of 150 m. No structural changes were implemented other than shortening or lengthening the tendons. Increasing the water depth did not affect Δ or $[M]$, and the effect on added mass was negligible. There was, however, a decrease in tendon stiffness due to the increased line length. The natural periods in surge/sway, heave, and yaw increased significantly when the water depth increased from 150 to 200 m, while the coupled pitch/bending natural periods remained relatively unchanged.

In lower sea states (ECs 1, 2, 6), $\sigma(\zeta_1)$ tended to increase with increasing water depth, but this trend was reversed as the wave height increased. Similarly, due to the decreased pitch stiffness, $\sigma(\zeta_5)$ increased with increasing water depth. The pitch motions in 200 m water depth were approximately 50% greater than in 100 m for several designs. The out-of-plane rotations, $\sigma(\zeta_4)$ and $\sigma(\zeta_6)$, behaved similarly to $\sigma(\zeta_1)$ with respect to water depth. These motions, which can contribute to increased power variability and wind turbine component fatigue, should be carefully examined, particularly for deeper water.

Despite the decrease in $\sigma(\zeta_5)$, $\sigma(M_{FA})$ increased 5–10% in the 100 m water depth. Inertia forces took up more of the wind and wave forcing in deeper water (200 m), while the structural reactions became more important in shallow water (100 m). The tendon tension was also fairly sensitive to the decrease in water depth: $\sigma(T_1)$ increased by approximately 20% for TLPWTs 2–5 in 100 m for ECs 3 and 4, while $\sigma(T_1)$ decreased by roughly 10% for TLPWT 1. On the other hand, $\sigma(T_1)$ did not change as much when the water depth was increased to 200 m. There was a 3–8% increase in $\sigma(T_1)$ for TLPWTs 1 and 2, a 2–3% decrease for TLPWTs 3–4, and a 3–5% decrease for TLPWT 5.

5.5.3. Pontoon radius

The effects of changing the pontoon radius by 20% depended on the baseline design. For the designs that depended primarily on the main column for displacement, an increase in rP shortened Tn_6 (due to the increased stiffness) and slightly shortened the coupled pitch-bending periods, with little effect on Tn_1. For designs that depended heavily on the pontoons for displacement, the increase in Δ often outpaced the increase in mass (particularly when the turbine and ballast mass were significant compared to the steel mass), resulting in higher pretension and shorter surge and yaw natural periods. Tn_3 was slightly longer due to the increase in mass and added mass with no change in the tendon axial stiffness.

Variations in rp had clear effects on the pitch motions, tendon tension, and out-of-plane rotations. Even for the designs that did not depend heavily on the pontoons for buoyancy, a significant reduction (25–40%) in pitch motion was achieved by increasing rp by 20%, while an even more significant (50–100%) increase in $\sigma(\zeta_5)$ occurred for a corresponding decrease in rp. The increase in rp also reduced $\sigma(T_1)$ by up to 20%, while a 20% decrease in pontoon radius resulted in up to a 50% increase in $\sigma(T_1)$. The out-of-plane motions were very sensitive to rp: the 20% increase in rp reduced $\sigma(\zeta_4)$ and $\sigma(\zeta_6)$ by 40–50%, indicating that this may be an effective method to improve the yaw stability.

The extent of the effects of varying rp on $\sigma(\zeta_1)$ and $\sigma(M_{FA})$ depended more on the TLPWT design and environmental conditions. The increased surge stiffness due to the buoyancy increase tended to reduce the surge motions in ECs 1, 2 and 6, but the benefit of the added stiffness decreased as the wave height and period increased. An increase in rp led to larger $\sigma(\zeta_1)$ in ECs 4 and 5 for almost all of the designs. The effects of rp on $\sigma(M_{FA})$ were small (less than 2%), even for TLPWTs 4 and 5, which had the largest changes in displacement (note that the remarks on TLPWT 5 do not consider a decrease in rp, only an increase).

5.5.4. Ballast fraction

An increase in the ballast fraction (BF) represents a larger quantity of concrete permanent ballast in the main column, without any change in geometry. That is, the mass and inertia increase and the pretension decreases. An increase in BF increased the system inertia and decreased the tendon pretension without affecting the hydrodynamic added mass, damping, and wave excitation forces. The system natural frequencies decreased when BF increased.

The effects of the changes in BF on the in-plane motions did not show a clear trend. An increase in ballast (and corresponding decrease in stiffness) resulted in up to 50% larger$\sigma(\zeta_1)$ – particularly for TLPWTs 1 and 2 – in the lower sea states, but reduced $\sigma(\zeta_1)$ for ECs 4 and 5. The ballast increase had variable effects on $\sigma(\zeta_5)$: increasing BF increased the pitch motions of TLPWT 1 and 5 the most, slightly increased $\sigma(\zeta_5)$ for TLPWT 2, and reduced $\sigma(\zeta_5)$ for TLPWT 3 and 4. Overall, the in-plane motions of the designs with the largest ballast fractions (1 and 5) were also most improved by decreasing BF by 0.1. Clearer trends in $\sigma(\zeta_4)$ and $\sigma(\zeta_6)$ were noted. The out-of-plane rotations increased significantly (up to 70% for TLPWT 1) when BF increased, indicating that increased stiffness is more effective than increased inertia in reducing $\sigma(\zeta_4)$ and $\sigma(\zeta_6)$.

Increasing BF resulted in a more distinct increase in $\sigma(T_1)$ and a small but distinct decrease in $\sigma(M_{FA})$. The reduction in $\sigma(T_1)$ may be related to the corresponding decrease in Tt: designs with larger pretension tended to have less tension variation. TLPWT 1 was most affected by the increase in BF with almost 80% increase in $\sigma(T_1)$ in ECs 4 and 5. The up to 8% decrease in $\sigma(M_{FA})$ associated with increased BF indicated that the increased mass allowed more of the environmental forces to be taken up by inertia. A reduction in BF gave corresponding decreases in $\sigma(T_1)$ and increases in$\sigma(M_{FA})$.

5.6. Design trends

By examining the results for all of the parametric variations, it was possible to identify some design trends that can be used to guide further TLPWT development.

The natural periods of the structure should be chosen carefully to improve the performance. For example, Fig. 10 indicates that $\sigma(\zeta_1)$ in operational conditions increased with longer surge natural periods, but decreased in the extreme condition. Tn_1did not have a large effect on T_1, but the designs with longer surge periods also tended to have decreased $\sigma(M_{FA})$.

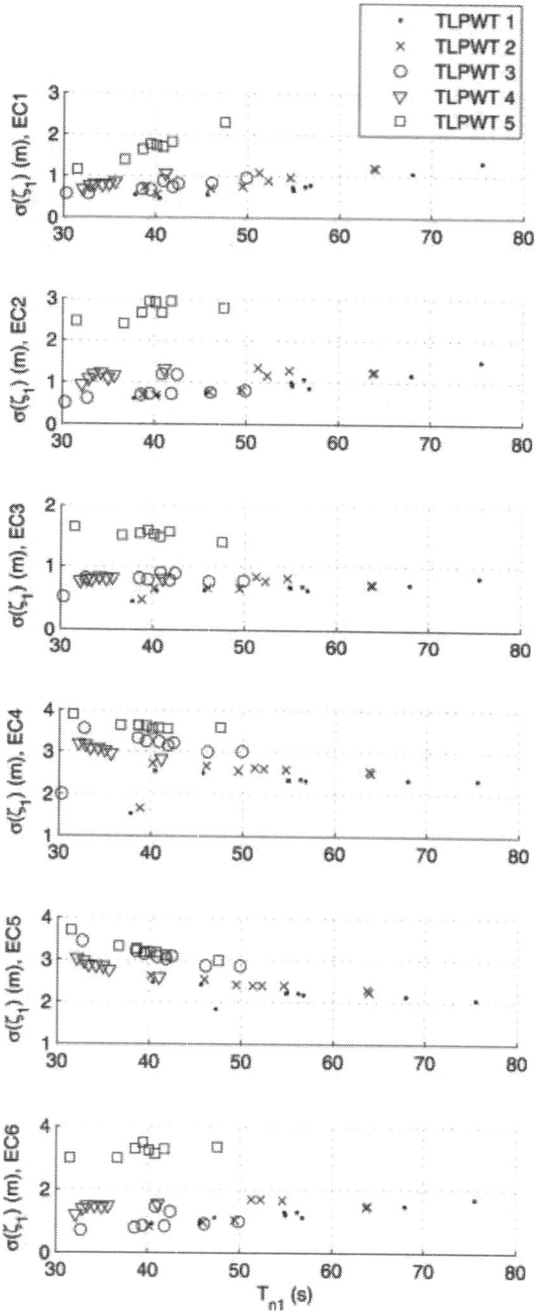

Figure 10. Effect of surge natural period on surge motions.

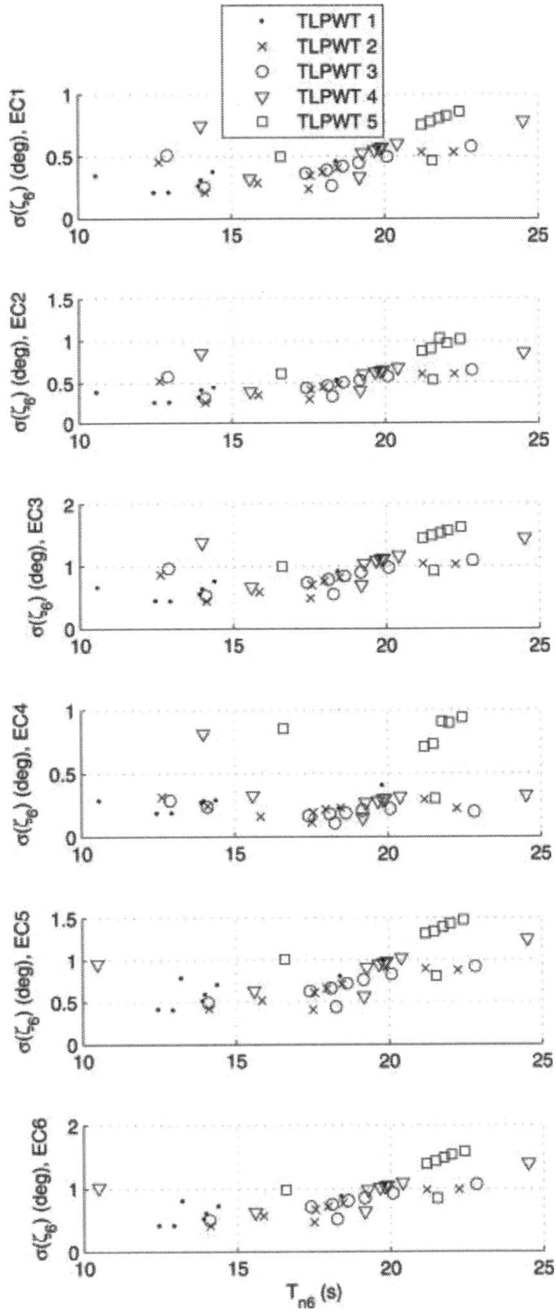

Figure 11. Effect of yaw natural period on yaw motions.

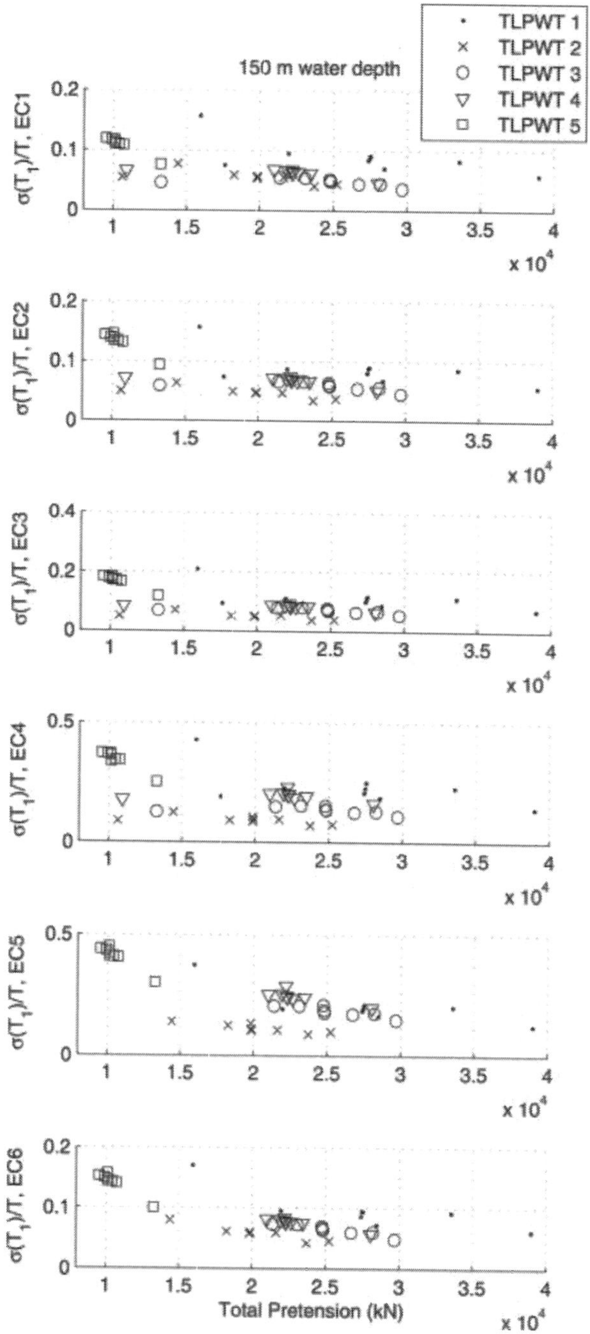

Figure 12. Effect of total pretension on variation in tendon tension.

Within the range of observed heave periods (0.4–0.9 s), there was no strong trend relating $\sigma(T_1)$ or $\sigma(M_{FA})$ to Tn_3. It may be more important to consider the relationship between the Tn_3 and Tn_5 in order to mitigate tendon loading, rather than tuning the Tn_3independently.

Reducing Tn_6 tended to decrease the yaw and roll motions, as shown in Fig. 11 for $\sigma(\zeta_6)$. Because the mean values of yaw and roll were nearly zero, changes in the Tn_6 primarily affected the standard deviations. This result agrees with the reduction in yaw motion when *rp* was increased. In terms of tower loading, however, Tn_6 did not appear to have a significant influence on M_{SS}.

The range of pitch/bending natural frequencies in this study was very small. A trend toward less pitch motion for longer Tn_5 was observed, but, except in the storm conditions EC 4 and 5, $\sigma(M_{FA})$ was not seen to be significantly reduced by increasing Tn_5.

In addition to natural frequencies, overall sizing considerations are a critical aspect of TLPWT design. The results did not show any important trends based on D_1 and D_2alone, but the choice of displacement and total pretension was important for the variation in T_1. As shown in Fig. 12, the tendon variation decreased as the total pretension increased, particularly when results are examined within each TLPWT design and its variations. Note that the results shown in Fig. 12 do not include depth variations. When the results were considered in terms of displacement rather than pretension, a reversal in the trend was noted for TLPWT 1, which has the largest displacement. This indicates that a smaller design with less permanent ballast may perform equally well in terms of tendon tension variation. On the other hand, the variation in M_{FA} was only weakly dependent on the displacement, and did not show a strong relation to the total pretension.

6. CONCLUSION

The current work examines design trends and considerations for single column TLPWTs. Because construction costs increase roughly linearly with displacement and pretension, the design goal should be to achieve the best performance with the least required displacement and lowest possible pretension. For all of the designs that met some minimum requirements, placing the wind turbine on the TLP platform did not have a large effect on the turbine blade loads or power output, but the bending moment variation at the tower base increased significantly compared to the land-based turbine. In general, the designs with the largest displacement showed the smallest motions and the designs with the highest pretension showed the least relative tension variation. Motions perpendicular to the wind and waves were found to be an important consideration, even in aligned wind-wave conditions. Roll motions, which were very lightly damped, may be excited by the time-dependent wind turbine and generator torque, and yaw motions may be excited both by roll coupling and by the time-varying aerodynamic loading. Due to the low damping, the out-of-plane rotations exceeded the pitch motions in some cases, but occurred at lower frequency.

The TLPWT responses were more dependent on changes in natural periods and stiffness than direct dependence on diameter, water depth, ballast fraction, or pontoon radius. For example, the surge natural period needed to be sufficiently long (>45 s) to reduce surge motions in storm conditions but not too long (<70 s) to see decreased performance in operational conditions, while decreasing the yaw natural period reduced the platform yaw motions. The wind turbine tower base bending moment was only weakly dependent on displacement and pitch/bending natural period.

Although the optimal TLPWT design has not been found, TLPWTs 3 and 4 with increased pontoon radius tended to show the best overall behavior considering platform motions, tendon tension, and turbine loads in the considered environmental conditions, and the roughly estimated construction cost. This study was very limited in its consideration of different environmental conditions, and additional simulations are required to more accurately predict fatigue and extreme events. Furthermore, the detailed design and structural integrity of the hull, especially at the pontoon attachment, was not investigated. The results of this study, which uses a coupled nonlinear code, may be used to develop more efficient analysis routines for optimization and improved design. Based on the current results, the authors suggest that a reasonable single column design with displacement between 3500 and 6500 m^3, three pontoons (for easier installation), and pontoon radius between 28 and 35 m may be able to support a 5 MW wind turbine and withstand harsh environmental conditions.

ACKNOWLEDGMENTS

Support for this work was provided by Statoil through an MIT-NTNU Gemini cooperative research project, and by the European Commission through the 7th Framework Programme (MARINA Platform – Marine Renewable Integrated Application Platform, Grant Agreement 241402). The authors are grateful to MARINTEK for support in the development of the coupled computer code, and to Steven Leverette of SBM Atlantia and Dagfinn Sveen of Olav Olsen for valuable discussions.

A. SIMPLE LINEAR APPROXIMATIONS

A.1. Added mass

Neglecting interaction between the components and end effects, 2-D added mass coefficients can be applied for each of the components based on Ref. [43]. For a cylinder with diameter D, the transverse added mass per unit length is given by Eq. (6).

$$a_t[D] = \rho \pi D^2/4 \tag{6}$$

The added mass per unit length for a square section with side length h is estimated as in Eq. (7).

$$a_t[h] = 4.754\rho\left(\frac{h}{2}\right)^2 \tag{7}$$

Eq. (8) shows the summation of the contributions to the surge added mass from the main column and pontoons. The contribution of the pontoons is summed based on the angle θabout the z-axis.

$$A_{11} \approx a_t[D_1](h_1 - b_t) + a_t[D_2]h_2 + \sum_{i=1}^{n_p} l_p a_t\left[h_p \text{ or } d_p\right]\cos^2\theta_i \tag{8}$$

The heave added mass due to the main column (Eq. (9)) is approximated as the added mass of a sphere with diameter equal to D_2, plus contributions due to the pontoons[43] and [26].

$$A_{33} \approx \rho\pi\frac{1}{12}D_2^3 + \sum_{i=1}^{n_p} l_p a_t\left[w_p \text{ or } d_p\right] \tag{9}$$

The added mass in pitch and coupled surge-pitch is computed by integrating the sectional added masses as in Eqs. (10) and (11).

$$A_{55} \approx a_t[D_1](H_1 - b_t)\left(\frac{1}{12}(H_1 - b_t)^2 + \left(\frac{1}{2}(H_1 - b_t)\right)^2\right) + a_t[D_2]H_2\left(\frac{1}{12}H_2^2 + \left(\frac{1}{2}(-2T + H_2)\right)^2\right)$$
$$+ \sum_{i=1}^{n_p} \cos^2\theta_i\left(\frac{1}{3}\left(r_p^3 - D_2^3\right)a_t\left[w_p \text{ or } d_p\right]\right) + \sum_{i=1}^{n_p} \sin^2\theta_i\left(z_s^2\left(r_p - \frac{D_2}{2}\right)a_t\left[h_p \text{ or } d_p\right]\right) \quad (10)$$

$$A_{51} \approx -\frac{1}{2}(H_1 - b_t)^2 a_t[D_1] - \frac{1}{2}\left(T^2 - (-T + H_2)^2\right)a_t[D_2]$$
$$+ \sum_{i=1}^{n_p} \sin^2\theta_i z_s\left(r_p - \frac{D_2}{2}\right)a_t\left[h_p \text{ or } d_p\right] \quad (11)$$

Similarly, the yaw added mass is computed by summing the integrated effects of the pontoons (Eq. (12)). No contribution to the yaw added mass from the central column is considered.

$$A_{66} \approx \sum_{i=1}^{n_p} a_t\left[w_p \text{ or } d_p\right] \frac{1}{3}\left(l_p^3 - \frac{D_2^3}{2}\right) \quad (12)$$

A.2. Hydrostatic stiffness

Considering the platform as a rigid body, the hydrostatic stiffness depends only on the water line geometry, overall center of buoyancy (zB) and overall center of gravity (zG). Eqs. (13) and (14) can be used to determine the nonzero terms in the hydrostatic stiffness matrix.

$$C_{33} = \rho g A_{wp} \quad (13)$$

$$C_{55} = C_{44} = \rho g I_{wp} + \rho g V z_B - M g z_G \tag{14}$$

A.3. Mooring system stiffness

The mooring system stiffness matrix due to nt lines at positions θj is approximated for small motions by assuming that the lines remain straight. Following for example[28] and [27]:

$$k_{11} = \frac{T_0}{l_0}, k_{33} = \frac{E_t A_t}{l_0} \tag{15}$$

$$K_{11} \approx \sum_{j=1}^{n_t} k_{11} \tag{16}$$

$$K_{33} \approx \sum_{j=1}^{n_t} k_{33} \tag{17}$$

$$K_{51} = K_{15} \approx \sum_{j=1}^{n_l} k_{11} z_s \tag{18}$$

$$K_{55} \approx \sum_{j=1}^{n_l} \left[k_{11} z_s^2 + k_{33} r_p^2 \right] \cos^2(\theta_j) \tag{19}$$

$$K_{66} \approx \sum_{j=1}^{n_l} k_{11} r_p^2. \tag{20}$$

For the given systems' symmetry, we also note that:

$$K_{22} = K_{11} \tag{21}$$

$$K_{24} = K_{42} = -K_{51} \tag{22}$$

$$K_{44} = K_{55}. \tag{23}$$

REFERENCES

1. Butterfield S, Musial W, Jonkman J, Sclavounous P and Wayman L. Engineering challenges for floating offshore wind turbines. In: 2005 Copenhagen offshore wind conference, no. NREL/CP-500-38776; 2005.
2. Robertson AN and Jonkman JM. Loads analysis of several offshore floating wind turbine concepts. In: 21st International Offshore and Polar Engineering conference, no. NREL/CP-5000-50539; 2011.
3. Matha D. Model development and loads analysis of an offshore wind turbine on a tension leg platform, with a comparison to other floating turbine concepts. Master's thesis. University of Colorado-Boulder: April, 2009.
4. Henderson AR, Argyriadis K, Nichos J and Langston D. Offshore wind turbines on TLPs – assessment of floating support structures for offshore wind farms in German waters. In: 10th German wind energy conference, Bremen, Germany: 2010.
5. Moon III WL, Nordstrom CJ. Tension leg platform turbine: a unique integration of mature technologies. In: Proceedings of the 16th offshore symposium. Texas Section of the Society of Naval Architects and Marine Engineers; 2010. p. A25–34.
6. Casale C, Lembo E, Serri L, Viani S. Preliminary design of a floating wind turbine support structure and relevant system cost assessment. Wind Engineering 2010;34(1):29–50.
7. Tracy C. Parametric design of floating wind turbines. Master's thesis. Massachusetts Institute of Technology; 2007.

8. Sclavounos P, Tracy C and Lee S. Floating offshore wind turbines: responses in a seastate; pareto optimal designs and economic assessment. Tech. rep. Department of Mechanical Engineering, Massachusetts Institute of Technology; 2007.

9. Crozier A. Design and dynamic modeling of the support structure for a 10 MW offshore wind turbine. Master's thesis. NTNU (Norwegian University of Science and Technology); June 2011.

10. Fulton G, Malcolm D, Elwany H, Stuart W, Moroz E and Dempster H. Semi-submersible platform and anchor foundation systems for wind turbine support. Tech. rep. NREL/SR-500-40282. National Renewable Energy Laboratory; 2007.

11. Sclavounos P, Lee S, DiPietro J, Potenza G, Caramuscio P and Michele GD. Floating offshore wind turbines: tension leg platform and taught leg buoy concepts supporting 3–5 MW wind turbines. In: European wind energy conference EWEC 2010, Warsaw, Poland: 2010.

12. Matha D, Fischer T, Kuhn M, Jonkman J. Model development and loads analysis of a wind turbine on a floating offshore tension leg platform. no. NREL/CP-500-46725; 2010.

13. Stewart G, Lackner M, Robertson A, Jonkman J and Goupee A. Calibration and validation of a fast floating wind turbine model of the deepcwind scaled tension-leg platform. In: 22nd International Offshore and Polar Engineering conference, no. NREL/CP-5000-54822; 2012.

14. Nihei Y, Matsuura M, Fujioka H and Suzuki H. An approach for the optimum design of TLP type offshore wind turbines. In: Proceedings of the 30th international conference on ocean, offshore, and arctic engineering, no. OMAE2011-50258; 2011.

15. Copple RW and Capanoglu C. Tension leg wind turbine conceptual design suitable for a wide range of water depths. In: 22nd international ocean and polar engineering conference; 2012.

16. Suzuki K, Yamaguchi H, Akase M, Imakita A, Ishihara T, Fukumoto Y, et al. Initial design of TLP for offshore wind farm. In: Renewable energy conference 2010, Yokohama, Japan: 2010.

17. Jonkman JM and Buhl ML Jr. FAST user's guide. Tech. rep. NREL/EL-500-38230. National Renewable Energy Laboratory; August 2005.

18. Winterstein SR, Ude TC, Kleiven G. Springing and slow-drift responses: predicted extremes and fatigue vs. simulation. In:

BOSS'94 Proceedings of the seventh international conference on the behaviour of offshore structures; 1994.

19. Chakrabarti S, editor. Handbook of offshore engineering. Elsevier Science; 2005.

20. Bae YH, Kim MH and Shin YS. Rotor-floater-mooring coupled dynamic analysis of mini TLP-type offshore floating wind turbines. In: The 29th international conference on ocean, offshore, and arctic engineering, no. OMAE2010-20555; 2010.

21. Bachynski EE and Moan T. Linear and nonlinear analysis of tension leg platform wind turbines. In: The 22nd international ocean and polar engineering conference, no. 2012-TPC-0629; 2012.

22. Jonkman J, Butterfield S, Musial W and Scott G. Definition of a 5-MW reference wind turbine for offshore system development. Tech. rep. NREL/TP-500-38060. National Renewable Energy Laboratory; February 2009.

23. Jonkman J. Definition of the floating system for Phase IV of OC3. Tech. rep. NREL/TLP-500-47535; 2010.

24. Kibbee SE, Leverette SJ, Davies KB and Matten RB. Morpeth SeaStar mini-TLP. In: Offshore technology conference, no. OTC 10855; 1999.

25. Leverette S. Personal communication; 2011.

26. Chakrabarti SK and Hanna SY. Added mass and damping of a TLP column model. In: Offshore technology conference, no. OTC 6406; 1990. p. 559–71.

27. Jain AK. Nonlinear coupled response of offshore tension leg platforms to regular wave forces. Ocean Engineering 1997; 24(7):577–92.

28. Faltinsen O. Sea loads on ships and offshore structures. Cambridge University Press; 1990.

29. American Petroleum Institute. Planning, designing, and constructing tension leg platforms (Edition 3). Tech. rep.; 2010.

30. MARINTEK. SIMO user's manual; 2011.

31. MARINTEK. RIFLEX user's manual; 2011.

32. Moriarty PJ and Hansen AC. AeroDyn theory manual. Tech. rep. NREL/TP-500-36881; 2005.

33. Luxcey N, Ormberg H and Passano E. Global analysis of a floating wind turbine using an aero-hydro-elastic numerical model. Part 2:

benchmark study. In: The 30th international conference on ocean, offshore, and arctic engineering, Rotterdam, the Netherlands, no. OMAE2011-50088; 2011. [34]

34. Ormberg H, Passano E and Luxcey N. Global analysis of a floating wind turbine using an aero-hydro-elastic model: part 1: code development and case study. In: The 30th international conference on ocean, offshore, and arctic engineering, Rotterdam, the Netherlands, no. OMAE2011-50114; 2011.

35. Ormberg H and Bachynski EE. Global analysis of floating wind turbines: code development, model sensitivity and benchmark study. In: The 22nd international ocean and polar engineering conference; 2012.

36. Johannessen K, Meling TS and Haver S. Joint distribution for wind and waves in the northern North Sea. In: The 11th international offshore and polar engineering conference & exhibition; 2001.

37. International Electrotechnical Commission. Wind turbines – part 3: design requirements for offshore wind turbines. Tech. rep. IEC61400-3; 2009.

38. Jonkman B. TurbSim user's guide: version 1.50. Tech. rep. NREL/TP-500-46198. National Renewable Energy Laboratory; September 2009.

39. International Electrotechnical Commission. Wind turbines: part 1: design requirements. Tech. rep. IEC61400-1:2005(E); 2005.

40. Bachmann H, Ammann WJ, Deischl F, Eisenmann J, Floegl I, Hirsch GH, et al. Vibration problems in structures. Birkhäuser Verlag; 1995.

41. Larsen TJ and Hanson TD. A method to avoid negative damped low frequent tower vibrations for a floating, pitch controlled wind turbine. Journal of Physics: Conference Series. In: The second conference on the science of making torque from wind, vol. 75; 2007.

42. Jonkman JM. Influence of control on the pitch damping of a floating wind turbine. In: Proceedings of the 2008 ASME wind energy symposium, no. NREL/CP-500-42589; 2008.

43. Newman JN. Marine hydrodynamics. The MIT Press; 1977.

4

CHAPTER

Characteristics of Tides in the Red Sea Region, a Numerical Model Study

Fawaz Madah, Roberto Mayerle, Gerd Bruss, Joaquim Bento

Coastal Research Laboratory-CORELAB, Research and Technology Centre West Coast (FTZ), University of Kiel, Kiel, Germany

ABSTRACT

In this work, a two-dimensional numerical model based on Delft3D modelling system was setup to study the tidal characteristics of the Red Sea. Besides that, analyses of available observed time series of surface elevations were carried out. Sensitivity analyses of the numerical model were carried out by testing different model parameters aiming at selecting optimal settings. The model performance was evaluated against available time series of surface elevation observations. RMS error was found to vary from 0.03 to 0.1 meter, while the ADM values range from 0.02 to 0.05 meter. On the whole, the model is able to reproduce the tidal wave in the Red Sea, reflecting a consistent level of agreement with the observations and previous works. The model results suggest that the semidiurnal tidal waves play a major role in the region except in the central part of the Red Sea where amphidromic system exists. Major semidiurnal and diurnal tidal constituents were computed to generate co-charts and form factor. The results have revealed that the distribution of the co-charts of the major semidiurnal constituents M_2, N_2, and S_2 show the existence of anticlockwise amphidromic system in the central part of the Red Sea at about 19.5°N, north of the Strait of Bab el Mandeb at 13.5°N and in the Gulf of Suez. The chart of the diurnal tidal constituent K_1 showed a single counterclockwise system in the southern part of the Red Sea centred around 15.5°N. The form factor chart shows the appearance of diurnal character in the central part of the basin and the northern end of the strait. The hydrodynamics patterns under spring and neap tidal conditions were also analysed (during flood and ebb conditions). Model results showed that currents generally are weak and strongest currents are present in the strait of Bab el Mandeb and Gulf of Suez.

Keywords: Red Sea, Delft3D Modelling System, Amphidromic System, Co-Charts

1. INTRODUCTION

Tides are the periodic motion in sea level resulting from the gravitational force of the moon and the sun and the centrifugal force generated as the earth rotates about the common centre of mass earth-moon-sun system. A basic knowledge of ocean dynamic particularly tides propagation at any place is a prerequisite to predict problems such as pollution, dispersion, sediment transport, etc. on coastal regions. Tidal currents influence the erosion processes occurring daily, and hence affect the nutrients budgets [1] [2] . The Red Sea is an elongated-shaped basin (semi-enclosed), extending from NNW to SSE (30°N to 12°N latitude and 32°E to 44°E longitude) over a distance of about 2000 km and separates the African and Asian landmass (Figure 1). The physical boundary of the Red Sea is delineated by the coastline of Egypt, Sudan, Eritrea and part of Djibouti in the western side and the coastline of Jordan, Saudi Arabia and Yemen on the eastern side. In the southern part of the basin, it links with the Gulf of Aden and the Indian Ocean through a narrow, shallow strait (26 km wide, 150 km long) known as Bab el Mandeb while in the north it consists of the so-called Gulf of Suez and Gulf of Aqaba [3] -[6] . The Red Sea experiences irregular bottom topography. The average depth is about 490 m while maximum depth > 2000 is found in the axial trough. The Red sea has a surface area of about 450,000 km^2. In the southern part of the basin, there are shallow shelves on both sides of the strait of Bab el Mandeb linking the Red Sea with the Gulf of Aden, the shallowest Sill known as Hanish Sill is located at the northern part of the Strait and the narrowest contraction is located at the Perim Narrows in the southern end of the strait [7] . The Red Sea contains a unique large marine ecosystem. Most of the coastline of the Red Sea is bordered by a coral reef system, thus tide and associated currents basically are important components for ecological and relevant environmental studies. On the other hand, the Red Sea is an important international shipping path linking eastern and southern Asia with the Middle East and Europe. Despite its socio-economic and environmental importance, there is very little published information regarding tidal characteristics in the region. The earlier works on tides were basically limited to observations

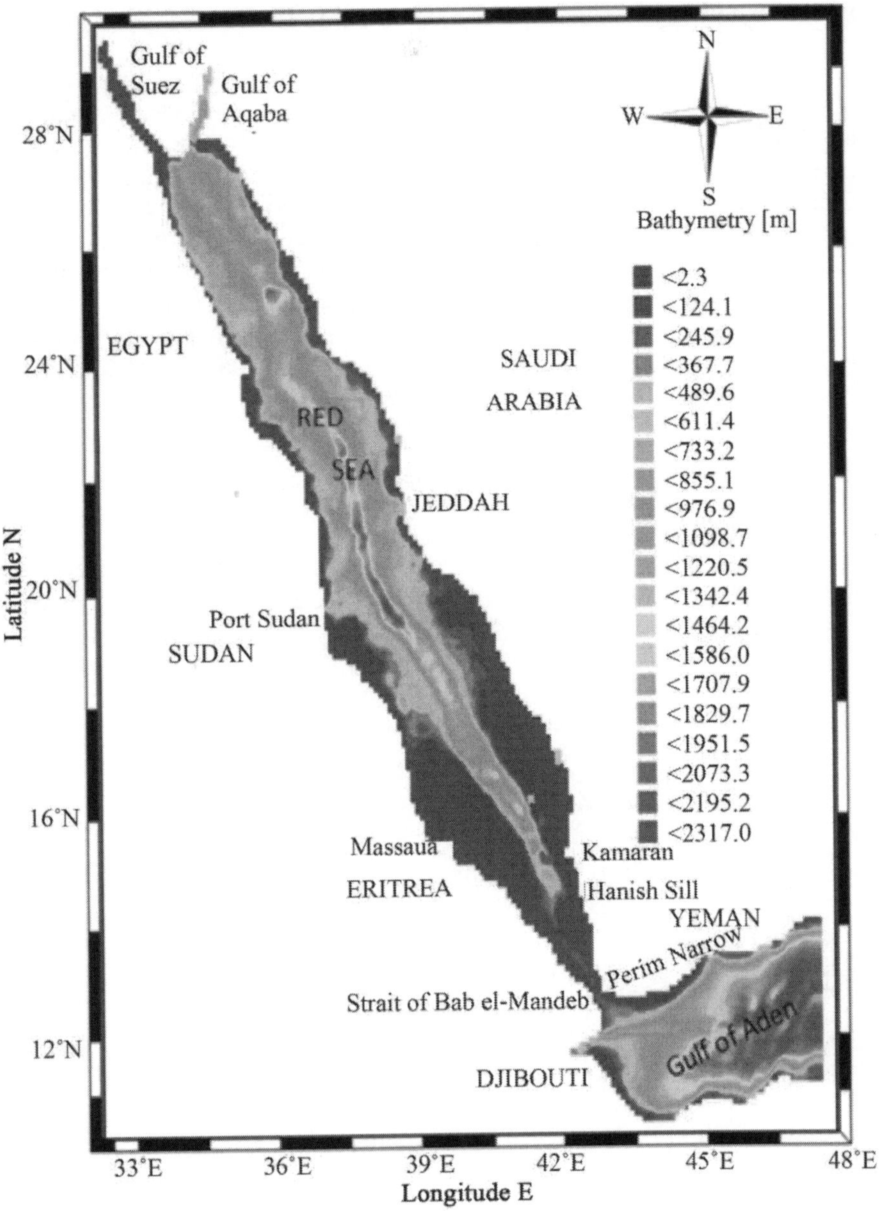

Figure 1. Geographic location of the Red Sea and the numerical bathymetry based from the global bathymetry datasets for the world ocean "GEBCO_08".

which are confined to the coastal region [6] [8] [9] . During the last century, a dynamical explanation of actual tides was supplied based on theoretical study. Studies on tides based on numerical modelling approach do not exist. Many researchers emphasize the necessity of numerical modelling approach to advance basic understanding of the region where there is lack of understanding the physical system regulating the marginal seas and Gulfs in the region [10] . Consequently, it is of interest to carry out a study of tides in the Red Sea using a modelling approach. Therefore, this research focuses on tidal characteristics in the Red Sea region by means of implementation and application of a numerical modelling method and by analysing its results. The specific model used to achieve this study is Delft3D modelling system, developed by Delft Hydraulics in the Netherlands. Besides that, available observations of surface elevations in the region were also considered in this study.

This paper is organized as follows. In section 1.1, a review of the study area is presented. In section 1.2, the available observations of surface elevation are analyzed. Section 2 introduces a brief description of the modelling system used in this investigation, followed by the setup of the Red Sea model (RS-Model) including processes involved in the setup and model performance. Section 3 presents the results of the numerical simulations including the tidal characteristics based on the distributions of co-range and co-tidal charts, form factor and hydrodynamic features. Section 4 throws light on the conclusion of this study.

1.1. General Review

A general description regarding the physical characteristics of the Red Sea such as wind conditions, water temperature, salinity, heat flux, etc, is found in the previous studies of [5] [11] -[19] . Generally, the climate of the Red Sea region is a typical desert and semi-desert climate characterized by high temperatures, low rainfall with an average of 109 mm/year. The northern part of the Red Sea is under the influences of the eastern Mediterranean weather system. South of 18° latitude including the Gulf of Aden region is mainly controlled by the Indian monsoon system, north-easterly monsoon (October to May) and a south-westerly monsoon (June to September). The evaporation rate in the Red Sea is very high compared to other oceans and seas. Fresh water is negligible due to low precipitation and absence of river discharge. As a result, highly saline conditions are found in the Red Sea (42 psu).

Very little information is found concerning tides in the literature. The tides in the Red Sea have been described by [20] based on a few analytical analyses carried out in order to determine the tides in the strait of Bab el Mandeb and the results of these analyses were used to explain tidal dynamics in the whole Red Sea basin. The conclusion drawn suggests that the tidal regime in the Red Sea is essentially co-oscillations with those of the Gulf of Aden, characterized by its low tidal range with semidiurnal characteristics. There is a difference of six hours between the time of high water in the north and in the south. This means that high water at the southern end of the Red Sea signals low water at the northern part and vice versa. The tidal range changes from north to south with greatest values at the two ends. In the north, at the entrance of Suez and Aqaba Gulfs the spring range is about 0.6 m while in the southern part of the Red Sea (at Massawa and Kamaran Island) is around 0.9 m. However, the range decreases towards the central part of the Red Sea near Jeddah and Port Sudan where amphidromic system (counter-clockwise) exists due to the M_2 tidal wave. There are also another two nodal zones that exist across the Red Sea with negligible tide occurring to the north of Bab el Mandeb and at the entrance of the Gulf of Suez. Generally, in places where the semidiurnal tides are very weak, diurnal character may appear. Nevertheless, it has been pointed out by [12] that there is disagreement on a complete explanation of the Red Sea tides.

Reference [9] analyzed time series of hourly sea level records for one year at two stations located in the central part of the Red Sea (Jeddah and Port Sudan). Their analysis showed that the tide at Jeddah can be classified as mixed type dominated by semidiurnal tide. On the other hand, their results indicated that the semidiurnal components (M_2, S_2) at Port Sudan are smaller in comparison with those at Jeddah and the amplitudes of the fortnightly and monthly components are larger. Based on the form ratio (F) value, the tide at Port Sudan was found to be of diurnal type. This was explained on the basis of the proximity of Port Sudan to the amphidromic point. Reference [6] performed harmonic analysis of hourly sea level records for the period (1992-1994) in the southern part of the Red Sea at JIZAN (Lat 16°53'N, Long 42°35'E). His analysis showed that the tide at JIZAN is dominated by the large amplitudes of the major semidiurnal constituents M_2, S_2, and N_2. Diurnal components on the other hand (O_1, K_1) were the smallest amplitudes among the others. Using the form ratio (F) to determine the character of the tide, it was found that the ratio was less than 0.25 indicating the type of semidiurnal tide.

Observational studies on water currents inside the Red Sea domain are very limited, and the patterns that emerged from the observations are significant northward currents close to the western coast in the southern part of the Red Sea during winter [21] and southward counter-current at depths greater than 250 m in the central part of the Red Sea in the open sea but less than 150 m near the coast of Port Sudan [22] ; a cyclonic gyre at the extreme north (north of 23° latitude) which penetrates up to 300 m depth and an anticyclonic gyre about 2° to the south [4] [14] [19] [22] [23] . These previous studies are basically attributing and discussing the importance of the wind effects and thermohaline forcing on the generated currents. In the past century, the area that attracted the researchers is the strait of Bab el Mandeb. As a result, most of the observational studies were carried out in the vicinity of Bab el Mandeb strait which were aimed at understanding the exchange between the surrounding water bodies. The results of these investigations suggest that there is a 2-layer exchange flow system in a winter period from October to May, with surface inflow and a deep outflow characterized by very saline (40 psu) Red Sea Outflow Water (RSOW). In a summer season (June-September) the system of 2-layer is replaced by a 3-layer flow system; shallow surface outflow, subsurface intrusion of Gulf of Aden Intermediate Water (GAIW) and a deep RSOW [7] [13] [24] -[26] .

1.2. Tidal Characteristics Based on Observations

This section presents the tidal characteristics in the Red Sea based on the tidal constituents produced from the harmonic analysis of the available tidal observations. During this research the available data are the tidal elevations in the region of the Red Sea from five tide gauges stations along the eastern coast of the Red Sea (JIZAN, JEDDAH, YANBU, RABIGH and DUBA) operated by Saudi ARAMCO Oil Company, three new tide gauges were installed in 2011 in the vicinity of JEDDAH coastal waters at the Coast Guard Stations of Obhor Creek, Gahaz and Saroom via the Faculty of Marine Sciences (King Abdulaziz University, Saudi Arabia) and Research and Technology Centre-CORELAB (University of Kiel, Germany) and data from one location on the western coast at Port SUDAN. In addition to that, time series of water level records from two additional stations located in the Gulf of Aden (at Aden and Djibouti) obtained from the Sea Level Centre at University of Hawaii (UHSLC) were made available (see Figure 3 for their locations). The purpose of using these data on the one hand is to perform tidal analysis and determine the major semidiurnal and diurnal constituents and on the other hand to be used for calibration and validation of the RS-Model. The length of the time series was one year at hourly intervals. Tidal analyses were performed using the World Tides program ver. 2009 [27] to compute the amplitudes and phases of the eight primary semidiurnal and diurnal tidal constituents (Q_1, O_1, P_1, K_1, N_2, M_2, S_2 and K_2). Examples of tidal amplitudes obtained from this analysis at some stations (ADEN, JIZAN, JEDDAH, RABIGH, DUBA and SUDAN) are shown in Figures 2(a)-(f). The figures obviously display a change in the tidal regime and range. It is clear that the tides of the Red Sea are relatively small and the magnitude varies according to the location. On the other hand, the magnitude of the tidal amplitude at Aden (Figure 2(a)) behaves differently, indicating different tidal regime characterized by a mixed type with diurnal and semidiurnal fluctuations. The figure shows that the major semidiurnal constituents are M_2 (principal lunar with a period of 12.42 hours), S_2 (principal solar with a period of 12.00 hours) and N_2 (larger elliptic lunar with a period of 12.66 hours) respectively, while the major diurnal tide is K_1 (principal luni-solar with a period of 23.93 hours). It is clear that the amplitudes of semidiurnal tidal constituents at SUDAN (Figure 2(f)) are smaller than diurnal constituents. Inside the Red Sea domain, the contribution of the semidiurnal constituents M_2, S_2 and N_2 for the total astronomical tide represents larger portion at location B, D and E. This indicates that the total astronomical tide is well represented by the major tidal constituents. On the other hand, the contribution of most other constituents on the total astronomical tide suggests that the constituents with smaller amplitude than the K_1 may be ignored.

2. MODEL SETUP

Only few observed data were found to be available in the area of interest during this research. Taking into consideration the length of the Red Sea, the availability of tide gauge data only at a few locations may not reflect a complete picture about the tidal characteristics in the region. However, considering the advantages of using a

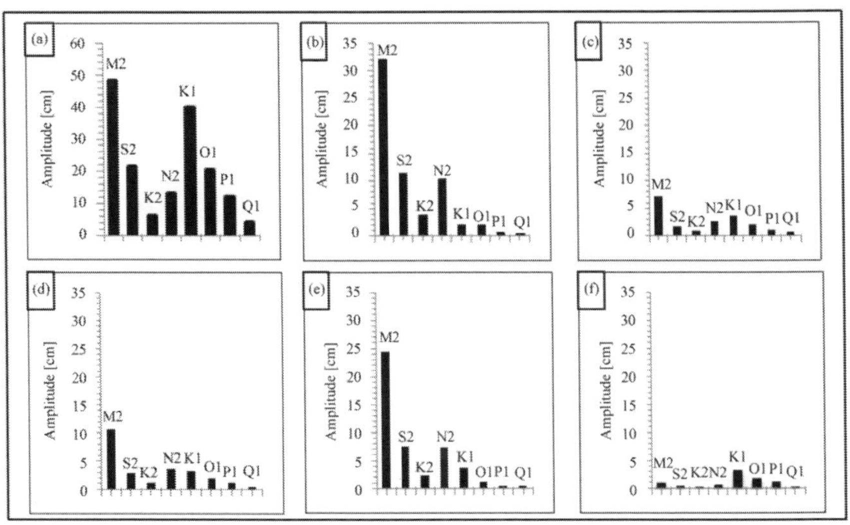

Figure 2. Amplitudes of semidiurnal and diurnal tidal constituents based on the tidal analysis, (a): ADEN; (b): JIZAN; (c): JEDDAH; (d): RABIGH; (e): DUBA; (f): SUDAN.

numerical modelling approach, the development of a numerical model and its application maybe a good alternative to carry out such study and provide a comprehensive picture about the tidal characteristics in the Red Sea. Two dimensional models, namely the depth-averaged or shallow water models, can be successfully applied for simulating tidal flow and flows of water bodies.

2.1. Model Description

In the present work, the Delft3D modelling system developed by WL|Delft Hydraulics in the Netherlands was used. The system is a fully integrated program which simulates the time and space variations of six phenomena namely hydrodynamics, wave interaction, water quality, morphology, sediment transport and ecology in addition to their interactions. The Delft3D-Flow is (the basic module in the system) a multi-dimensional (2D or 3D) hydrodynamic model which aims to solve non-steady flows and transport phenomena resulting from tidal and meteorological forcing including the effects of density differences (due to temperature and salinity distribution) on a rectilinear or curvilinear boundary fitted grid. It can be used to simulate flows in shallow seas, coastal areas, estuaries, lagoons, river and lakes. The equations of the system are derived from the three dimensional Navier Stokes equations for an incompressible free surface flow, under the shallow water and the Boussinesq assumption [28] . The system of equations consists of the horizontal equations of motion, the continuity equation and the transport equations for conservative constituents. The numerical solution of the equations is obtained by discretizing the equations in space using an explicit finite difference scheme over a uniform staggered Arakawa-C grid [29] and in time using Alternating Differential Implicit (ADI) technique. The ADI method divides one integration time step into two phases; each phase consists of half a time step [2] [30] . Since the solution is implicit, numerical stability is not limited by the time step, however, the accuracy of the flow decreases with increasing time step from accurate solutions when the Courant number is less than $4\sqrt{2}$ [31] .

2.2. Model Domain and Bathymetry

After gathering the input data needed for setting up the model including the bathymetry of simulation area and boundary condition, Delft-RGFGRID program was utilized to construct the grid.Figure 3 shows the resulting

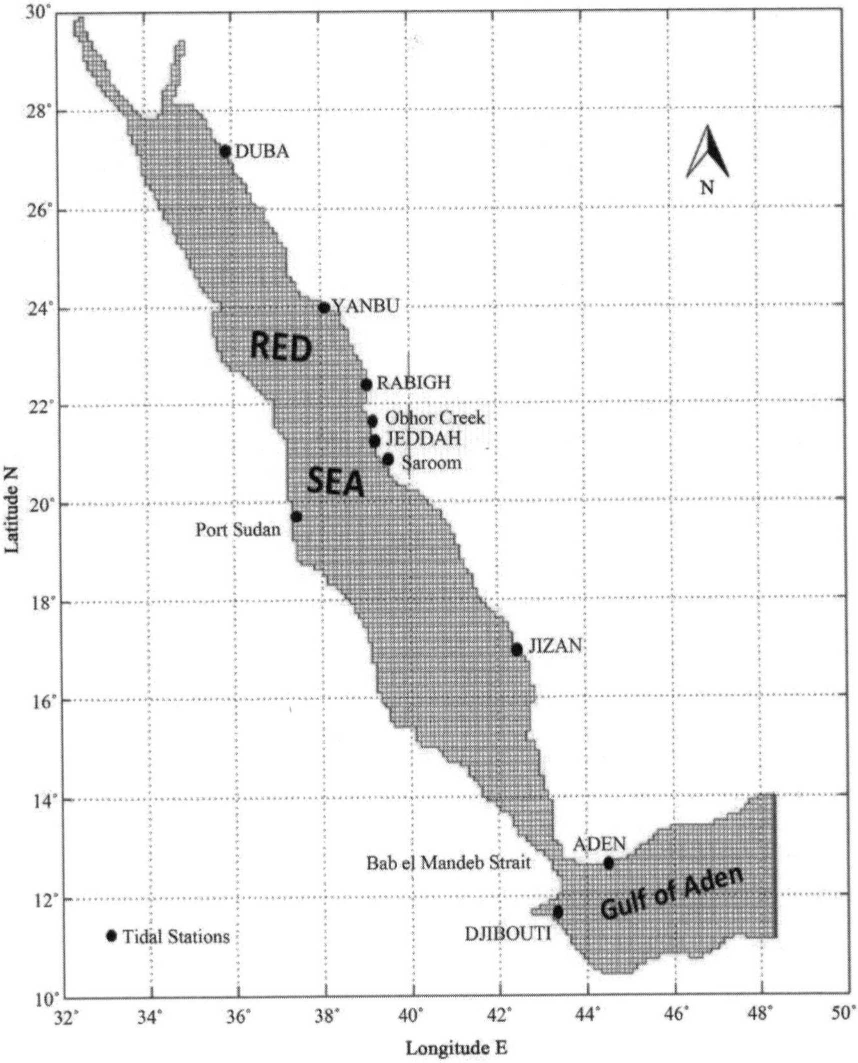

Figure 3. Computational grid distribution of the Red Sea model (RS-Model), red line represents the open boundary and green colour represents closed boundaries, [Black dots indicate the locations of the tidal gauge stations used in the model calibration and validation].

computational grid of the RS-Model where green colour represents the closed boundary while the open boundary is marked by red colour. The figure also includes the locations of the available surface elevation observations used for calibration and validation processes. The RS-Model domain covers the entire Red Sea and extends to the eastern part of the Gulf of Aden (48°E). The model domain was schematized on a rectangular mesh with a uniform grid spacing of 5 km horizontally. The total number of elements grid used in the computation consists of 63,434 grid cells. The coordinates system of the grid is spherical and therefore the variation of the Coriolis force is determined in the latitude direction.

The bathymetric data of the RS-Model were sourced from the global bathymetry dataset for the world ocean "GEBCO_08 Grid" at a 30 arc-second horizontal resolution (~1 km) (The GEBCO_08 Grid, version 20090202, http://www.gebco.net/). Data for the RS-Model domain were extracted and converted into the appropriate file format. Figure 1 shows the resulting bathymetric map based from this data. Due to insignificant artificial contact between the Red Sea and the Mediterranean via Suez Canal, the model has only one open boundary which was set at the Gulf of Aden at longitude (48°E) and divided into 25 segments. Due to unavailability of measurements near the open ocean boundary, tidal amplitudes and phases of the main semidiurnal and diurnal tidal constituents (Q_1, O_1, P_1, K_1, N_2, M_2, S_2 and K_2) extracted from the global ocean tidal model TPXO 7.2 on 1/4 × 1/4 degree resolution global grid were prescribed at boundary cells and linearly interpolated. A free slip condition was applied on the closed boundaries with a zero value of normal velocity. A time step of 60 seconds was set in all simulations to insure the stability and accuracy of the model. Initial conditions of water levels were set equal to zero in all simulations. As a result, the equilibrium state in terms of water levels is reached after certain time; therefore, the first four days of the simulation were discarded. Values of water density and gravitational acceleration were set 1025 kg·m^{-3} and 9.81 ms^{-2}, respectively.

2.3. Tuning the RS-Model

Since the approximation, simplification and the assumptions are applied in the numerical solution, which leads to differences between the model results and the observations, it is necessary to test the model inputs to insure that the model reproduce accurately the processes which were observed in reality [32] . Accordingly, various sensitivity analyses were carried out to analyze the effect of variations of different parameters on the hydrodynamic model results. This includes boundary condition, eddy viscosity, bottom roughness, mesh resolution, and time step. An early test of the numerical simulation was carried out to check the open boundary conditions. It was found that the model produces similar trend of variations of simulated and observed water level, however, discrepancies related to phase lag conditions were observed. Therefore, adjustments to the open boundary conditions were considered. Three different grid resolutions were tested in the purpose of selecting optimal grid spacing. This analysis revealed that the model is sensitive to grid resolution. Once a mesh grid was adopted, several computational scenarios for different time steps were carried out. This analysis revealed that the model is insensitive to time steps. However, a time step of 1 minute was adopted for stability and accuracy of the numerical model. Bottom roughness is one of the most important parameters in the hydrodynamic models. The Che'zy coefficient depends on the bottom roughness and the height of the water column. Sensitivity tests of Che'zy coefficient on the model results were carried out by testing three spatially and temporally uniform roughness values starting from 45, 65 and 85 $m^{1/2}/s$, respectively. It was found that using lower or higher value from the initial default value (65 $m^{1/2}/s$) given by Delft3D-Flow system produce insignificant influences on the model results. Therefore, a constant temporal and spatial value of 65 $m^{1/2}/s$ was adopted and applied throughout the model domain. Another important parameter is horizontal eddy viscosity (HEV) due to its role in defining the turbulence mixing in the water. A number of simulations were carried out to test the influence of HEV on the model outputs. This analysis revealed that surface elevation was not influenced by changing the HEV parameter. Therefore, a constant eddy viscosity of 1 m^2/s (default value of Delft3D-Flow system) was applied for the simulations. Table 1 lists the optimal parameters used for the RS-Model.

2.4. Model Validation

As part of quality control, validation of the RS-Model was considered in order to confirm the reliability and the quality of the model settings. This was done by considering another period different than the one used in the sensitivity analysis or in the calibration processes. RS-Model was validated against the available surface elevation data (mentioned previously in section 1.2) throughout the model domain. Beside the qualitative comparisons, model performance was evaluated at each location statistically by determining the mean absolute difference (MAD) and root mean square difference (RMSD). Comparisons between observed and simulated surface elevations at some of the stations are shown in Figures 4(a)-(d). These time series cover a period representing spring and neap tidal cycle. It is observed that the model predictions and observations of water levels are in very good agreement. The model is reproducing large scale tidal surface elevations within the Red Sea region and clearly provides a reasonable prediction of both the phase and amplitude of the spring and neap tide at each station. On the other hand, the model qualitatively captures major diurnal and semidiurnal features of the tides in the stations of Gulf of Aden adequately. To quantify the differences between the time series obtained from simulations and the measurements, a statistical analysis was carried out. This was done by computing the root mean square error (RMSE) and absolute mean difference (AMD) between the predicted and observed water levels. Statistical analysis showed that RMS error for all stations in the Red Sea region including the stations located in the Gulf of Aden was found to vary from 0.03 to 0.1 meter, while the ADM values range from 0.02 to 0.05 meter.

Table 1. The optimal parameters used for the Red Sea model.

Parameters	Value	Unit
Total Number of grids	63,434	-
Grid Size	5	km
Time Step	1	Minute
Bottom Roughness: Che'zy Coefficient	65	$m^{1/2}/s$
Horizontal Eddy Viscosity HEV	1	m^2/s

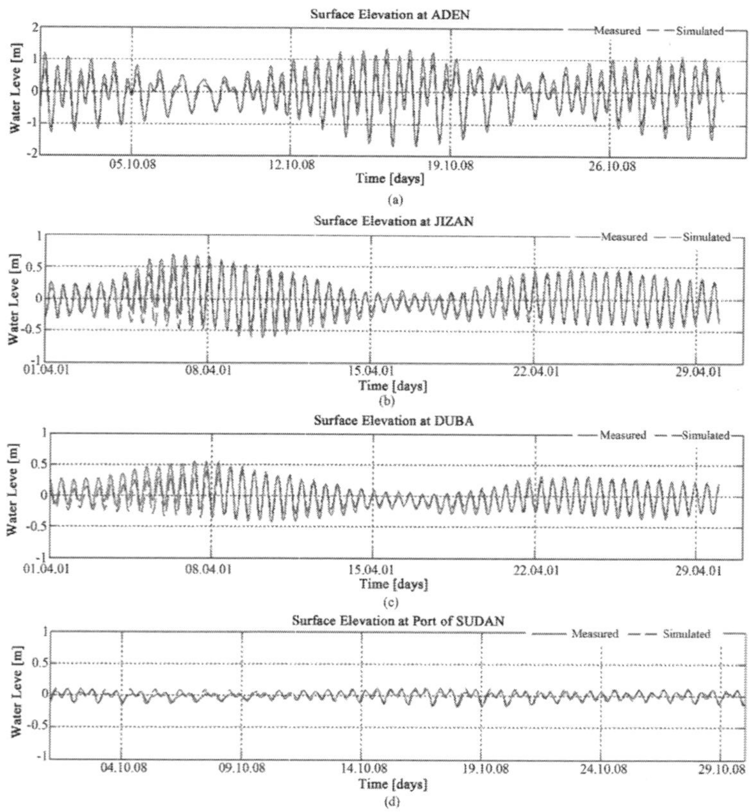

Figure 4. Comparison between the simulated and measured surface water elevation, (a): ADEN, (b): JIZAN, (c): DUBA, (d): SUDAN.

Table 2. Observed and simulated amplitude [m] and phase [degrees] of the major constituents including their differences for some coastal stations.

Constituents	M2	S2	N2	K1	Station ID	Latitude	Longitude
Amp_{mod}	0.425	0.20	0.108	0.18			
Amp_{obs}	0.484	0.215	0.130	0.4			
Dif_Amp	−0.059	−0.015	−0.022	−0.22	ADEN	12°78'	44°97'
Pha_{mod}	136.51	148.12	134.25	347.64			
Pha_{obs}	134.76	157.09	152.52	341.41			
Dif_Pha	1.75	−8.97	−18.27	6.23			
Amp_{mod}	0.307	0.116	0.081	0.01			
Amp_{obs}	0.321	0.104	0.104	0.019			
Dif_Amp	−0.014	0.012	−0.023	−0.009	JIZAN	16°52'	42°33'
Pha_{mod}	37.34	65.11	6.714	165.3			
Pha_{obs}	34.0	68.0	8.00	160.7			
Dif_Pha	3.34	−2.89	−1.286	4.6			
Amp_{mod}	0.048	0.013	0.015	0.019			
Amp_{obs}	0.071	0.015	0.024	0.033			
Dif_Amp	−0.023	−0.004	−0.009	−0.014	JEDDAH	21°31'	39°08'
Pha_{mod}	179.6	202.2	145.8	216.0			
Pha_{obs}	202.6	244.2	164.6	207.4			
Dif_Pha	−23	−42	−18.8	8.6			
Amp_{mod}	0.114	0.032	0.040	0.018			
Amp_{obs}	0.152	0.45	0.048	0.029			
Dif_Amp	−0.038	−0.418	−0.008	−0.011			
Pha_{mod}	107.92	248.36	190.94	232.27	YANBU	23°06'	38°13'
Pha_{obs}	1.9	235.3	261.7	197.7			
Dif_Pha	106.02	13.06	−70.76	34.57			
Amp_{mod}	0.224	0.082	0.060	0.028			
Amp_{obs}	0.245	0.075	0.074	0.038			
Dif_Amp	−0.021	0.007	−0.014	−0.01	DUBA	27°19'	35°43'
Pha_{mod}	210	239.7	178.8	219.7			
Pha_{obs}	219	229.0	267.7	155.3			
Dif_Pha	−9	10.7	−88.9	64.4			
Amp_{mod}	0.031	0.08	0.012	0.019			
Amp_{obs}	0.010	0.010	0.007	0.033			
Dif_Amp	0.021	0.07	0.005	−0.014	SUDAN	19°63'	37°11'
Pha_{mod}	245.2	243.6	274.4	157.6			
Pha_{obs}	204	219.2	351.8	144.0			
Dif_Pha	41.2	24.4	−77.4	13.6			

Amp_{mod} and Amp_{obs} represent the amplitudes obtained from model and observed data, respectively; Pha_{mod} and Pha_{obs} represent phases obtained frc model and observed data, respectively; Dif_Amp and Dif_Pha represent the amplitudes and phase's differences, respectively.

Simulated tidal constituents are also compared with the tidal constituents obtained from the observation data. The model was run for period of one year aiming at performing tidal analysis of the simulated time series of water level. For this purpose, the World Tides program ver. 2009 [27] was used. As mentioned earlier (section 1.2) the major semidiurnal constituents are M_2, S_2 and N_2while major diurnal tide is K_1. The differences between amplitudes and phase of the major semidiurnal and diurnal tidal constituents were calculated for both simulated and observed data and are tabulated in Table 2. At all stations, the tidal amplitudes of M_2, S_2, N_2 and K_1 derived from the model agree reasonably with the values determined from the observations. The major discrepancies were observed in relation to phase lag. The larger difference computed for the phase of the tidal constituents was observed for M_2 and N_2 tide respectively at YANBU and SUDAN stations compared to the observed phase, while S_2 and K_1 tide on an average compared to M_2 and N_2 tide showing minor discrepancies. On the other hand, discrepancies of the tidal components at ADEN and JIZAN are minor compared to other stations, reflecting a good agreement between the model and observations. This indicates that the tidal constituents prescribed along the open sea boundary are not the major source of the errors observed in YANBU and SUDAN. The discrepancies could be attributed to the model bathymetry. The performance of a numerical model depends highly upon accurate representation of the sea bottom levels. The bathymetry of the model was basically based on a deep- water data which does not include detailed bathymetry for shallow shelf waters. Therefore, such major discrepancies are expected at YANBU and SUDAN stations. On the whole, regardless of theses discrepancies there is a satisfactory agreement for both diurnal and semidiurnal constituents and the general results have shown that the model reproduces the tidal wave propagation in the Red Sea with a good accuracy.

3. TIDAL CHARACTERISTICS BASED ON MODEL RESULTS

This section presents the tidal characteristics in the Red Sea based on the tidal constituents generated from the harmonic analysis of the model results. Two numerical simulations were carried out to describe the tidal characteristics in the Red Sea. In the first scenario, simulations of major tidal constituents (M_2, S_2, N_2, and K_1) were made for the purpose of generating Co-tidal charts and also the form factor (section 3.1 and 3.2). Thus, separate simulations were carried out of each individual constituent to generate Co-tidal and Co-range charts for the Red Sea. The M_2 tide was prescribed and linearly interpolated to each grid point along the open boundary of the model. The simulated M_2 tidal elevation was stored at hourly intervals at all grid points in the model domain. The computed amplitudes and phase from simulated M_2 tide at each grid point were used to generate the Co-range and Co-tidal charts respectively. The same procedures were applied for simulations of other constituents (N_2, S_2 and K_1).

In the second simulation, the model was forced on the ocean open boundary with the amplitudes and phases of the primary semidiurnal and diurnal constituents (Q_1, O_1, P_1, K_1, N_2, M_2, S_2 and K_2). This simulation was carried out to reproduce the behaviour of the hydrodynamic features in the Red Sea at moments of extreme conditions, ebb and flood phases during spring and neap tidal cycles (section 3.3).

3.1. Co-Range and Co-Tidal Charts

To represent the propagation of tides in the Red Sea, the so-called co-range and co-tidal charts are used. Co-tidal and co-amplitude charts of the semidiurnal constituents M_2, N_2, and S_2 are shown in Figures 5(a)-(c) respectively. In general, the dominant feature of the M_2, N_2, and S_2 tide is the existence of anticlockwise amphidromic systems in the central part of the Red Sea close to 20°N, northern end of the strait of Bab el Mandeb at 13.5°N and the Gulf of Suez. The co-tidal and co-range chart of the diurnal constituent K_1 are displayed in Figures 5(d). The chart shows only a single anticlockwise amphidromic system in the southern part of the Red Sea centred around 15.5°N. These results are consistent with the description of tides in the Red Sea given by Defant [20] .

3.1.1. Semidiurnal Tidal Wave

Shown in Figure 5(a) are the co-amplitude and co-phase distributions for the M_2 tide. The former is given in centimetres and the latter is expressed in degrees. In general, the amplitudes of M_2, $S_2$and N_2 waves show similar behaviour in the Red Sea basin. It is obvious from the figure that M_2amplitude is relatively high in the Gulf of Aden (50 to 60 cm) however, as the tidal wave flow into the strait of Bab el Mandeb; the wave speed rapidly decreases along the strait direction due to the narrow connection. Further north, the decrease continues to minimum value as shown in the co-amplitude chart of Figure (5-a-right panel), then M_2 tidal amplitude becomes 5 cm or less in the strait of Bab el Mandeb and tends to increase north-westward. Defant [20] assumed that this minimum is associated with the amphidromic system generated due to M_2 tidal wave. At the northern end of the strait, M_2 tidal amplitude increases to 20 cm and then expands at both sides eastern and western coasts (15°N) in the southern part of the Red Sea reaching about 35 cm. The M_2 amplitude then tends to decrease from the southern part of the Red Sea toward north-northwest while reaching the central part where amphidromic system exists and it begins to propagate and increase again towards the northern part of the Red Sea. For S_2 tide, again the

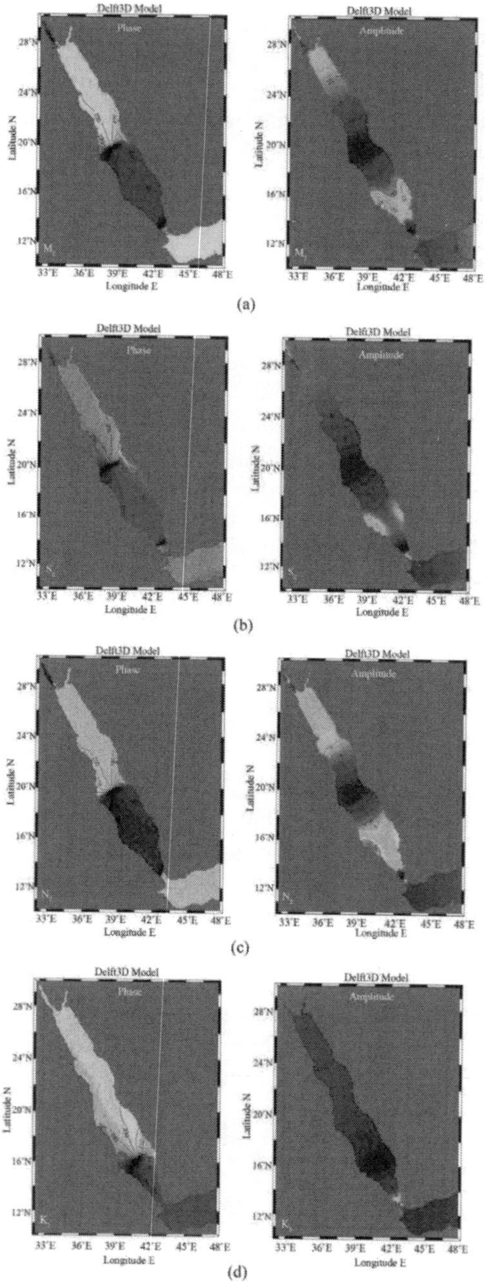

Figure 5. Distribution of amplitudes and phases of tidal components in the Red Sea using the RS-Model, (left panel) co-tidal lines and (right panel) co-range lines. (a): M_2; (b): S_2; (c): N_2 and (d): K_1. Amplitudes are given in centimetres and phases in degrees.

amplitude of S_2 tide is relatively high in the Gulf of Aden. The amplitude of S_2 tide is low (less than half) compared to the M_2 tide and its variation in the Red Sea is similar to that of M_2 tide as shown in Figure (5-b-right panel). These results are due to the influence of bathymetry (the strait morphology) in combination with the amphidromic system and results in decline in the tidal amplitude. The S_2 tide then propagates northward from the northern end of Bab el Mandeb strait to the inner area of the south of the Red Sea and tends to decrease (16° N) up to the amphidromic system. Similar to M_2 tide, the S_2 amplitude then propagates and increases gradually towards north-northwest. N_2 tide is similar to S_2 tide as illustrated in Figure 5(c). The amplitude of N_2 tidal wave decreases from the southern part of the Red sea toward the central part (amphidromic system) and tends to increase toward north-north west.

3.1.2. Diurnal Tidal Wave

The co-range and co-tidal charts of K_1 are depicted in Figure 5(d). The dominant feature of the phase distribution is the existence of a single anticlockwise amphidromic system located in the southern part of the Red Sea centred around 15.5°N. The distribution of the co-amplitude reflects higher amplitudes at the Gulf of Aden. However, in the strait of Bab el Mandeb, the amplitude of K_1as shown in the figure (right panel) is reduced to 10 cm and tends to decrease gradually towards the north along the strait. Contrary to those of semidiurnal constituents M_2, S_2 and N_2, K_1 tidal amplitude decrease gradually toward the inner area of the southern part of the Red Sea where amphidromic system is formed. The co-range chart of K_1 tide shows a slight increase in amplitude above 17°N towards middle part of the Red Sea but the amplitude is very small. Tidal wave of O_1(not shown) also includes similar features of K_1 tide.

3.2. Form Factor

According to Pugh [33] , the relative importance of the semidiurnal and diurnal tidal constituents can be determined based on a form factor. Therefore, it was considered the form factor for the entire Red Sea based on the computed amplitudes using the following expression [33] [34] :

$$F = \frac{O_1 + K_1}{M_2 + S_2}$$

where O_1, K_1, M_2 and S_2 are the amplitudes of the correspondent constituents. Shown in Figure 6 is the form

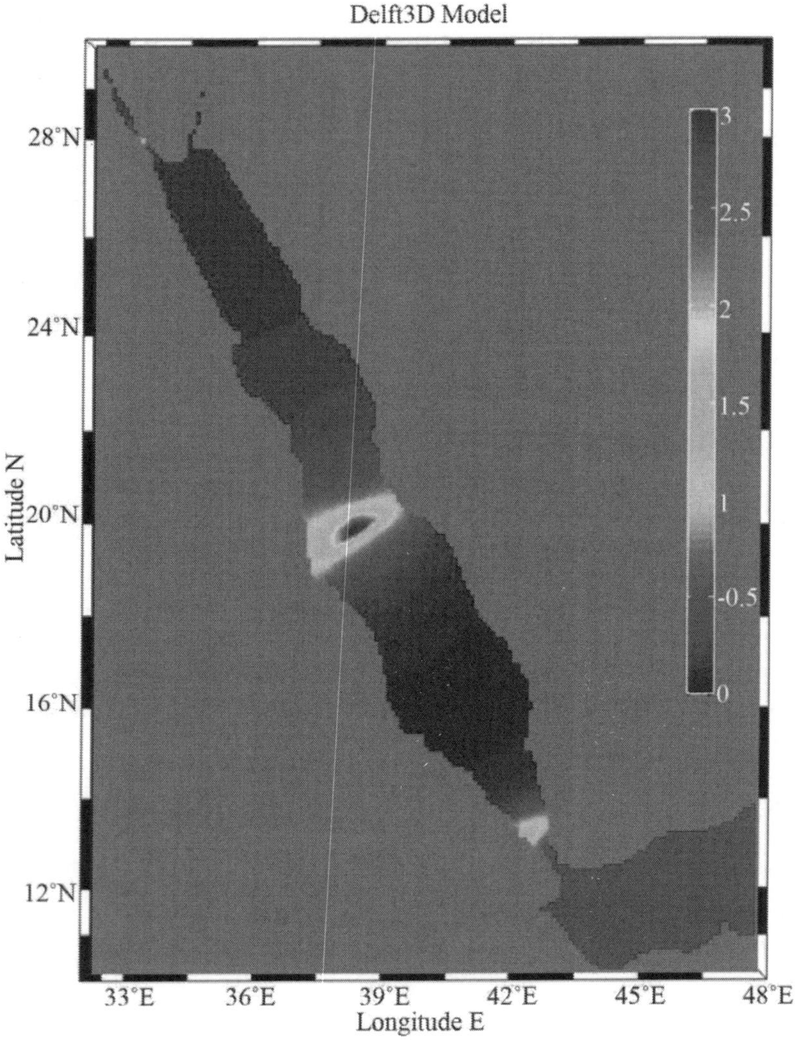

Figure 6. Form Factor distribution in the Red Sea based on calculated modelled diurnal (O_1, K_1) and semidiurnal (M_2, S_2) amplitudes.

factor distribution in the Red Sea basin. Looking at the figure, it is confirmed that the type of the tide in the Red Sea is essentially semidiurnal particularly in the southern and northern portion, as the form factor is less than 0.25 [33] [34] in all grid points of the computational domain except the central part of the Red Sea and northern part of the strait. Assessment of this figure has reflected that the relation between diurnal and semidiurnal constituents is to be seen in the central part of the Red Sea and is not constant in the entire Red Sea. Therefore, the diurnal tidal pattern is stronger in the central part of the Red Sea indicating mixed type mainly dominant with semidiurnal on the eastern coast compared with diurnal tide in the western coast. The results conducted by Sultan et al., [34] for the area of Port Sudan and Jeddah are in agreement with this work where diurnal character was observed to be larger close to the SUDAN coast. The figure also shows that diurnal character appears at the northern end of Bab el Mandeb strait around 13°N.

3.3. Hydrodynamic Features

This section presents the patterns of the tidal currents in the Red Sea based on the numerical model. Analysis of tidal currents that is generated by a combination of the primary constituents (Q_1, O_1, P_1, K_1, N_2, M_2, S_2 and K_2) under flood condition and ebb condition during the spring and neap tide were made. The simulated tide-induced current flooding and ebbing patterns in spring and neap tide are displayed in Figures 7(a)-(b). The highest velocities, in the range of 0.5 m/s were observed in Bab el Mandeb strait during flood tide during spring condition. However, the tidal currents are varying depending on the geometry in combination with ocean currents of flood and ebb tides. The snapshot of the current velocity field at the flood condition during spring tide is shown in Figure (7-a-left panel). At this particular time, the maximum current magnitudes reaches up to 0.5 and 0.3 m/s are observed in the vicinity of Bab el Mandeb strait and Gulf of Suez respectively. Analysis of ADCP data carried out in the Strait of Bab el Mandeb by [24] showed that the speed in the upper layer ranges between 0.4 - 0.6 m/s while for the lower layer is 0.8 m/s. The model results showed that current magnitude in the range of 0.1 m/s is prevalent inside the Red Sea. In the inner part of the southern portion of the Red Sea (16°N), current speeds become stronger at the vicinity of shallow shelves on both sides. On the other hand, during neap tide (Figure 7(b)) maximum currents magnitude in order of 0.25 m/s are observed in the strait of Bab el Mandeb during flood condition. During ebb condition, it was observed that velocities between 0.05 and 0.1 prevail inside the Red Sea domain.

Figure 8 and Figure 9 display distribution of the simulated surface elevations and currents during spring tides in the Red Sea, produced by a combination of the primary constituents (Q_1, O_1, P_1, K_1, N_2, M_2, S_2 and K_2). As it can be seen, the spring tides in the Red Sea are characterized by high water in the southern part of the basin whiles the northern part by low water and vice versa during ebb condition. Maximum elevations are observed in the southern part particularly over the shallow shelves, they are 0.5 m and 0.6 m close to eastern coast and western coast, respectively. The figure also reflects an increase in surface gradients along the Gulf of

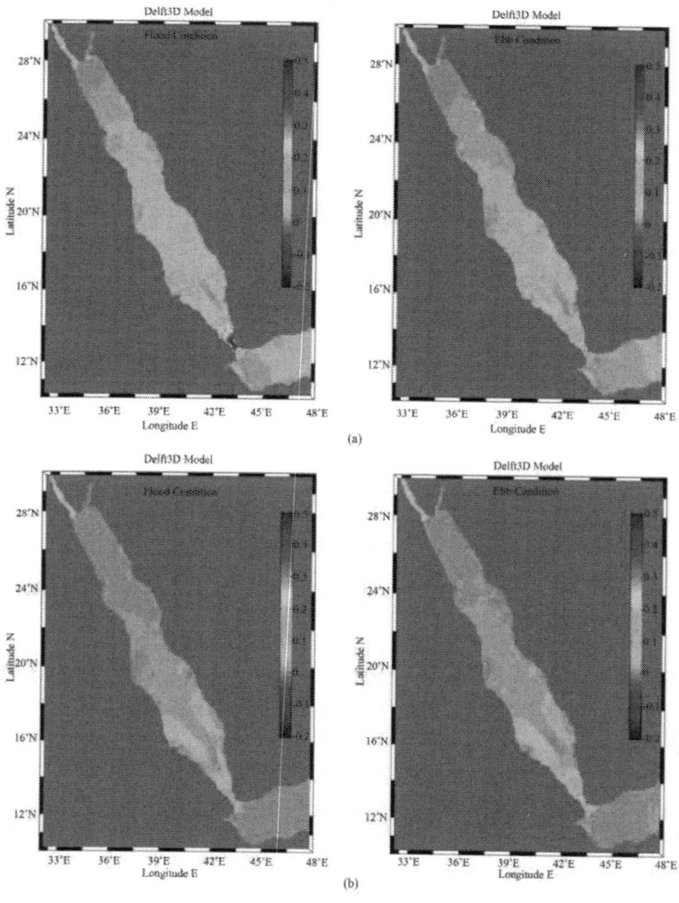

Figure 7. Snapshot of the model showing the current velocity field (m/s) for the flood condition (left panel) and for the ebb condition (right panel) (depicted by colour size) in the Red Sea during (a) spring and (b) neap tide respectively.

Suez. Tidal currents in the Red Sea are very weak with an average speed less than 0.1 m/s (Figure 9). The highest currents occur at the area where the strait is narrowing its land-water boundary and along the Gulf of Suez. The model results indicate that the direction of the currents in the Red Sea includes some variability. The currents during flood condition are directed northward however, the direction of the currents in the strait and Gulf of Suez is opposite. During ebb condition, the currents from the north propagate towards the southern part, however, in the strait and Gulf of Suez the direction is opposite.

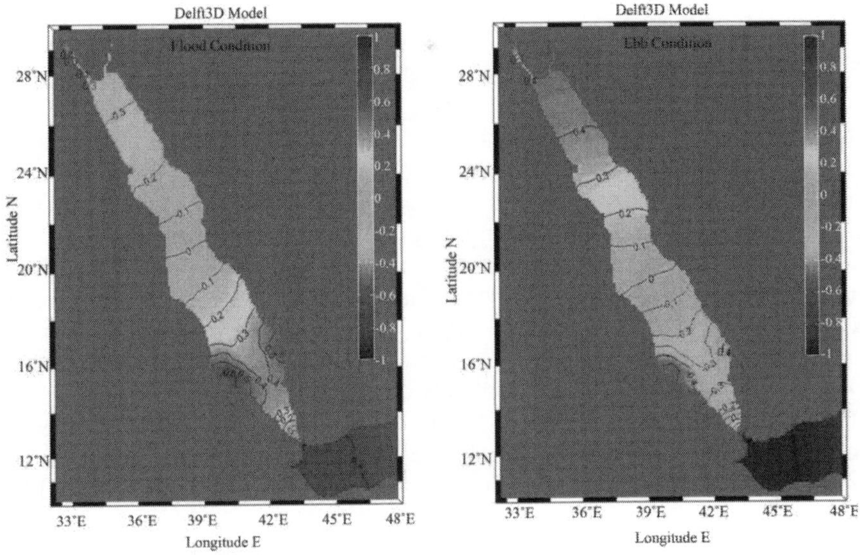

Figure 8. Distribution of the spring surface elevation (m) in the Red Sea using the RS-Model, (left panel) flood condition and (right panel) ebb condition.

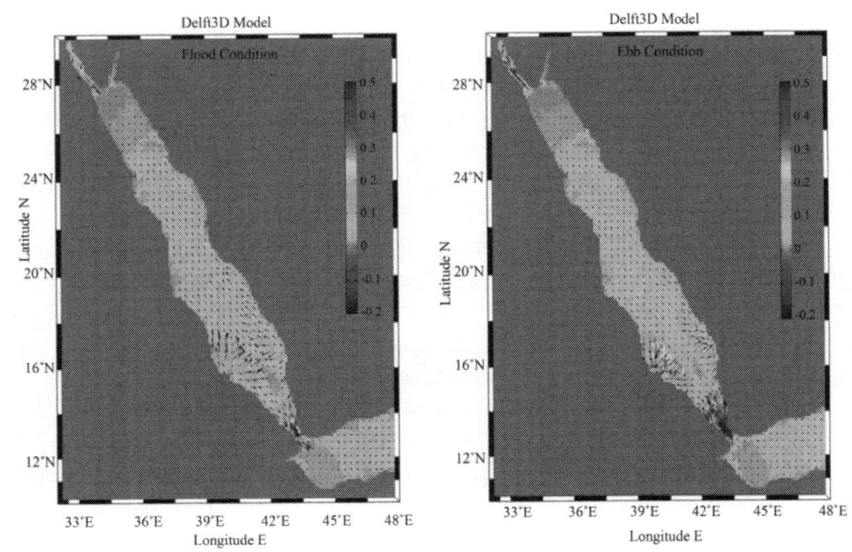

Figure 9. Snapshot of the model showing the current velocity field (m/s) for the flood condition (left panel) and for the ebb condition (right panel) (depicted by colour size and vector direction) in the Red Sea during spring tide.

4. CONCLUSIONS

Previous studies of tides in the Red Sea were limited to observations, in addition to a few analytical analyses carried out to determine the tides in the strait of Bab el Mandeb and the results of these analyses were used to explain the tidal dynamic for the whole Red Sea. Considering the advantages of numerical modelling, the tidal characteristics in the Red Sea region were studied using numerical model based on Delft3D modelling system. The model was run in two-dimensional mode. The model domain covers the entire Red Sea and includes part of the Gulf of Aden with open sea boundary located at 48°E. The optimal settings were determined based on sensitivity analyses. The model was validated against measured surface elevation at several locations in the model domain. The model is able to reproduce the tidal wave in the Red Sea, reflecting a consistent level of agreement with previous work and field data. At all stations considered in the validation, the tidal amplitudes of M_2, S_2, N_2 and K_1 derived from the model agree reasonably with the values determined from the observations. However, the major discrepancies were observed in relation to phase conditions at stations located near the region of amphidromic point. The major source of these errors could be attributed to the bathymetry of the model. Such errors are expected because a small error in position of the amphidrome can cause large phase errors [35] .

Based on the model results, the dominant feature of the M_2, N_2, and S_2 tide is the existence of the amphidromic systems (anti-clockwise) in the central part of the Red Sea at about 20°N, north the strait of Bab el Mandeb at 13.5°N and in the Gulf of Suez. The distribution of the co-phase of K_1tide showed only a single anticlockwise amphidromic system exists in the southern part of the Red Sea centred around 15.5°N. The amphidromic systems of the semidiurnal and diurnal constituents together in the strait of Bab el Mandeb suggest the tides include some characteristics of standing waves. Model results of amplitudes and form factor proved that tides in the Red Sea are dominated by the major semidiurnal constituents M_2, S_2, and N_2. However, diurnal character appeared in the central part of the Red Sea and northern part of the strait. In term of tidal currents, model results showed that tidal currents in the Red Sea are weak except near the Red Sea entrance where maximum velocity was observed to be 0.5 m/s. In summary, additional observations of water levels along the eastern coastline of the Red Sea are needed for more validation of the model since only one station was used in this study. In addition to that, measurements of current velocities are needed for validation purposes. Nevertheless, the model is useful and can be used to generate boundary conditions for local models in the Red Sea.

ACKNOWLEDGEMENTS

Fawaz Madah, Roberto Mayerle, Gerd Bruss, Joaquim Bento The authors would like to thank ARAMCO Oil Company, Department of marine physics at Faculty of Marine Sciences at King Abdulaziz University, Saudi Arabia, Research and Technology Centre-CORELAB at university of Kiel, Germany in particular Dr. José Manuel Fernández Jaramillo and Sea Level Centre at University of Hawaii for supplying the water level data. The acknowledgement is expressed to Prof. Dr. Fazal Ahmad Chaudhry for the valuable scientific suggestions on the earlier version of the manuscript.

REFERENCES

1. Canhanga, S. and Dias, J.M. (2005) Tidal characteristics of Maputo Bay, Mozambique. Journal of Marine Systems, 58, 83-97.
2. Kitheka, J.U., Ohowa, B.O., Mwashote, B.M., Shimbira, W.M., Mwaluma, J.M. and Kazungu, J.M. (1996) Water Circulation Dynamics, Water Column Nutrients and Plankton Produtivivity in a Well-Flushed Tropical Bay in Kenya. Journal of Sea Research, 35, 257-268.
3. Al Barakati, A.M. (2012) The Flushing Time of an Environmentally Sensitive, Yanbu Lagoon along the Eastern Red Sea Coast. International Journal of Science and Technology (IJST), 1, 53-58.
4. Clifford, M., Horton, C., Schmitz, J. and Kantha, L.H. (1997) An Oceanographic Nowcast/Forecast System for the Red Sea. Journal of Geophysical Research, 102, 25101-25122.
5. Patzert, W.C. (1974) Wind-Induced Reversal in Red Sea Circulation. Deep Sea Research, 21, 109-121.
6. Saad, M.A. (1997) Seasonal Flucuation of Mean Sea Level at Jizan, Red Sea. Journal of Coastal Research, 13, 1166- 1172.
7. Smeed, D. (2004) Exchange through the Bab el Mandeb. Deep-Sea Research, 51, 455-474.
8. Monismith, S.G. and Genin, A. (2004) Tides and Sea Level in the Gulf of Aqaba (Eilat). Journal of Geophysical Research, 109, C04015.

9. Sultan, S.A.R., Ahmad, F. and El-Hassan, A. (1995) Seasonal Variations of the Sea Level in the Central Part of the Red Sea. Esturine, Coastal and Shelf Science, 40, 1-8.

10. William, E.J., Gregg, A.J., John, C.K., Steven, P.M. and Mike, C. (1999) Arabian Marginal Seas and Gulfs, Report of a Workshop Held at Stennis Space Centre, Mississippi, University of Miami RSMAS, Technical Report 2000-01.

11. Ahmad, F. and Sultan, S.A.R. (1987) On the Heat Balance Terms in the Central Region of the Red Sea. Deep-Sea Research, 10, 1757-1760.

12. Edwards, A.J. and Head, S.M. (1987) Red Sea, Key Environments Series. Pergamon Press, Oxford, 42-78.

13. Maillard, C. and Soliman, G. (1986) Hydrography of the Red Sea and Exchanges with the Indian Ocean in Summer. Oceanologica Acta, 9, 249-269.

14. Morcos, S.A. (1970) Physical and Chemical Oceanography of the Red Sea. Oceanography and Marine Biology, Annual Review, 8, 73-202.

15. PERSGA (2006) The State of the Marine Environment of the Red Sea and Gulf of Aden (Regional Organization for the Conservation of the Environment of the Red Sea and Gulf of Aden). Report for the Red Sea and Gulf of Aden.

16. Quadfasel, D. and Baudner, H. (1993) Gyre-Scale Circulation Cells in the Red Sea. Oceanologica Acta, 16, 221-229.

17. Sheppard, C., Price, A. and Roberts, C. (1992) Marine Ecology of the Arabian Region: Patterns and Processes in Extreme Tropical Environments. Academic Press, London.

18. Sofianos, S.S., Johns, W.E. and Murray, S.P. (2002) Heat and Freshwater Budgets in the Red Sea from Direct Observations at Bab el Mandeb. Deep-Sea Research Part II, 49, 1323-1340.

19. Sofianos, S.S. and Johns, W.E. (2007) Observations of the Summer Red Sea Circulation. Journal of Geophysical Research, 112.

20. Defant, A. (1960) Physical Oceanography. Pergamon Press, Oxford.

21. Vercelli, F. (1927) The Hydrographic Survey of the R. N. Amrairaglio Magnaghi in the Red Sea. Annual Hydrographic, 2, 1-290.

22. Vercelli, E. (1931) Nuove richerche sulli correnti marine nel Mar Rosso. Annali Idrografici, XII, 12, 1-74.

23. Maillard, C. (1974) Eaux intermediaires et formation d'eau profonde en Mer Rouge. In: L'oceanographie physique de la Mer Rouge, Centre National Pour l'Exploitation des Oceans, Paris, 105-133.

24. Murray, S.P. and Johns, W. (1997) Direct Observations of Seasonal Exchange through the Bab El Mandeb Strait. Geophysical Research Letters, 24, 2557-2560.

25. Siedler, G. (1968) Schichtungs und Bewegungsverhaltnisse am Sudausgang des Roten Meeres. Meteor Forschungsergeb, 4, 1-67.

26. Morcos, S.A. and Soliman, G.F. (1974) Circulation and Deep Water Formation in the Northern Red Sea in Winter (Based on R/V Mabahiss Sections, January-February, 1935). In: L'oceanographie physique de la Mer Rouge, Centre National Pour l'Exploitation des Oceans, Paris, 91-103.

27. Boon, J.D. (2004) Secrets of the Tide: Tide and Tidal Current Analysis and Predictions, Storm Surges and Sea Level Trends. Horwood Publishing, Chichester, 212 p.

28. Roelvink, J.A. and Van Banning, G.K.F.M. (1994) Design and Development of DELFT3D and Application to Coastal Morphodynamics. In: Verwey, A., Minns, A.W., Babovic, V. and Maksimovic, C., Eds., Hydroinformatics, Balkema, Rotterdam, 451-456.

29. Arakawa, A. and Lamb, V.R. (1977) Computational Design of the Basic Dynamical Process of the UCLA General Circulation Model. Methods Computational Physics, 17, 173-265.

30. Stelling, G. and Leendertse, I. (1991) Approximation of Convective Processes by Cyclic ADI Methods. In: Spaulding, M.L., et al., Eds., Proceedings of the 11th International Conference on Estuarine and Coastal Modeling, American Society of Civil Engineers, Reston, VA, 771-782.

31. Stelling, G. (1984) On the Construction of Computational Methods for Shallow Water Flow Problems. Ph.D. Thesis, Rijkswaterstaat Communication Series No. 35, Rijkswaterstaat, The Hague.

32. Palacio, C., Mayerle, R., Toro, M. and Jiménez, N. (2005) Modelling of Flow in a Tidal Flat Area in the South-Eastern German Bight. Die Küste, Heft 69, 141-174.

33. Pugh, D. (1987) Tides, Surges and Mean Sea Level: A Handbook for Engineers and Scientists. John Wiley & Sons, Chichester, 472 p.

34. Pugh, D. (2004) Changing Sea Levels-Effects of Tides, Weather and Climate. Cambridge University Press, Cambridge, 265 p.

35. Jarosz, E., Blain, C.A., Murray, S.P. and Inoue, M. (2005) Barotropic Tides in the Bab el Mandab Strait—Numerical Simulations. Continental Shelf Research, 25, 1225-1247.

5

CHAPTER

Estimating Sea Bottom Shapes for Grounding Damage Calculations

Otto-Ville Edvard Sormunen [a,*], Antti Castren� [b], Jani Romanoff [a], Pentti Kujala [a]

[a] *Aalto University School of Engineering, Department of Applied Mechanics, Marine Technology, Otakaari 3, P.O. Box 12200, 02150 Espoo, Finland*

[b] *Finnish Transport Agency, Department of Information, Opastinsilta 12 A, 00520 Helsinki, Finland*

ABSTRACT

Groundings are among the most common and serious maritime accidents. The shape of the sea bottom is one of the important factors that determine the extent of grounding damage on ships, including loss of watertightness. This paper presents a four-step methodology for mathematically analyzing sea bottom shapes, where individual peaks are identified and isolated from larger datasets. To these individual peaks mathematical models are fitted and the goodness of the fit is evaluated.

The aim is to develop mathematical rock models that can be used in grounding damage analysis. The method is applied to bottom topography data from the two busiest tanker harbors in Finland. As potential mathematical rock models common assumptions found in literature as well as new suggestions are used.

It was observed that rocks vary in shape and size locally as well as between different geographical areas. This highlights the need to have rock models that are flexible enough to model a wide range of sizes and shapes so that the local conditions can be taken into account. This includes modeling asymmetrical shapes, which is not taken into account in the current models. The analysis results suggest using a binormal function to describe the bottom shape: This model gives overall better goodness-of-fit test results than the cone and polynomial models. In the data the binormal model can model well the also the bottom shapes where the polynomial and conical models had a good fit in statistical terms. Although the paper discusses data from the Gulf of Finland, the framework for analyzing bottom shapes is expected to be applicable to other sea areas as well.

Keywords: Groundings; Ship damage; Sea bottom shape; Gulf of Finland

1. INTRODUCTION

Groundings are among most common and serious maritime accidents. As grounding happens, significant parameters affecting the resulting damage are e.g. ship kinetic energy, loading condition and structural solutions (e.g. double bottom, compartments), the angle of attack to the ground and tides, and the properties of ground itself (e.g. shoal, rock). The relation between kinetic energy, structural solutions, the resulting damage and the residual strength of hull girder has been investigated by several authors using idealized bottom shapes such as cones and parabolas [1], [2], [3], [4], [8], [9], [10], [14],[15], [16], [17], [18], [21], [22], [23], [33], [34], [35], [36], [37], [39] and [40].

However, as stated by several authors (e.g. Refs. [1], [11], [24] and [34]) bottom shapes have central role in ship grounding event and therefore need further investigation. As an example of this [1] presented a comparative study on a single ship grounding on rock, reef and shoal. While sharp shape cuts the plating easily, the damage might be limited. As the size of indenter increases, the damage naturally spreads out for larger area in double bottom. Even if there is no hull fracture, the overall damage may harm the global strength of the hull girder [35], [1] and [5]. From a risk analysis perspective this has significant implications; e.g. on oil tankers a widespread damage might result in no oil spill if it is absorbed by the outer hull whereas a narrow but deep cut that penetrates the inner hull might result in a significant oil spill.

Another important issue is the grounding depth. Currently it is assumed to be either triangularly [13] and [30] or uniformly distributed [13]. In a triangular distribution the theoretical grounding depth range is between sea level ($z = 0$ m) and vessel draft. The most likely grounding depth is so that the sea bottom barely scrapes the ship bottom and the least likely grounding depth is at 0 m depth. [13] created the distributions based on expert opinions, not on empirical data. Grounding reports usually do not specify the actual grounding depth, thus there is a need for further research.

Tikka et al. [30] state that "Distributions for speeds, obstruction depths, and tides were based on the data from charts, NOAA[1], USCG[2] offices, and pilots," but the exact data or methodology behind arriving at triangular distribution for the grounding depth is not specified.

This paper describes a systematic approach to identify individual peaks from the sea bottom data pool and how the individual peaks can be modeled mathematically for grounding damage calculations. The paper also aims at testing the assumptions regarding grounding depth and the current conical rock shape assumption using bottom topographical data from selected Finnish harbors. Having correct assumptions regarding these factors is important for the damage estimation in the existing grounding models.

1.1. Current rock shape assumptions in literature

Currently the bottom shape during grounding is assumed to be a symmetrical cone-like object with varying diameter and slope. Most commonly the rock is assumed to be a blunted cone, see e.g. Refs. [15], [21] and [26]. There is some variations in how the cone tip looks like, i.e. whether it is simply cut-off or rounded, see Fig. 1. Another rock model found in literature is the polynomial rock used by Ref. [9]; who use a 2nd order polynomial equation to describe the rock, see Fig. 1. The equation is

$$z = -\frac{y^2}{a}$$

where a is the form parameter of the rock for the 4 different rock sizes illustrated inFig. 1.

The sea bottom shapes of [1] are relatively similar to the ones in Fig. 1. Other fields have also dealt with geological formations; The geometry of mountainous landscapes (e.g. Refs. [20], [25] and [7]) and rocks (e.g. Refs. [27], [38], [19] and [29] has been analyzed in literature. [28] present a method for quantitatively and qualitatively analyzing similarities between surfaces including peak regions. Other relevant publications include [28] alpha peak region analysis, sand grain shape parameter classification of [6] and automatic mound- [32] and glacier valley [25] detection. However, the methods found in literature do not cover extensively this paper's research question; no comprehensive method to identify, isolate, classify and mathematically model individual peaks from larger topographical data was found.

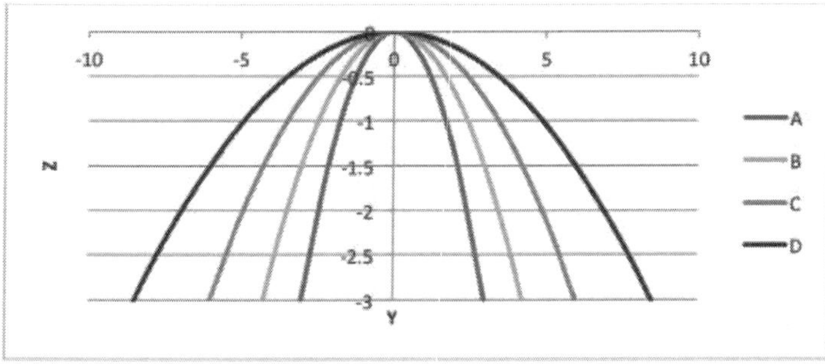

Rock	a [m]
A	3
B	6
C	12
D	24

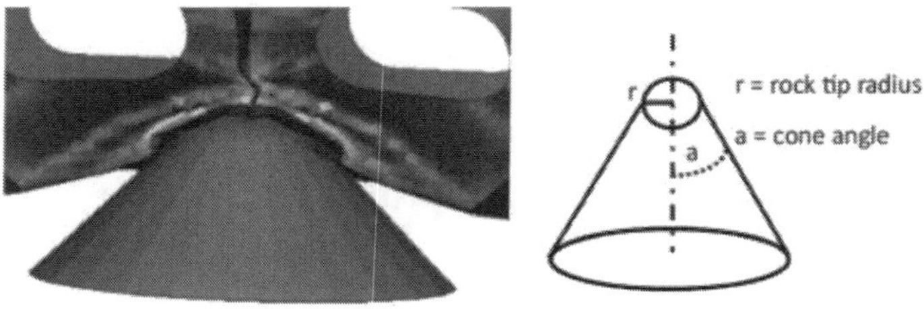

Figure 1. Rocks and their parameters from Ref. [21] (bottom left), [26] (bottom right) and [9] (top and table).

2. METHODOLOGY

The bottom shape analysis based on this data is carried out using the following four-step approach:

Step 1:Define relevant sea area and find bottom data

Step 2:Identify individual peaks from the larger bottom formations

Step 3:Define the cut-off method for the individual peaks

Step 4:Fit mathematical models to the data and evaluate the models

The steps are described in detail as follows:

Step 1:Define relevant sea area and find bottom data

In this case, two fareways from the Gulf of Finland was selected for analysis. On these fareways all rock formations that follow the following criteria are further selected for analysis:

1a)Rock formation is in the immediate vicinity of the fareway so that ships can possibly ground there.

1b)Sufficiently high resolution data is available of the location, e.g. data from multibeam echo soundings.

1c)The formation has at least one bottom shape apex which is at a depth of no less than*m* meters and no deeper than the fareway nominal draft clearance.

Restriction c) is due to the draft limitations of the survey vessels, limiting availability of data from shallow waters as they cannot be covered from all sides even with side-sweeping sonar equipment. In this case *m* is approximately 5 m. This restriction is illustrated Fig. 2.

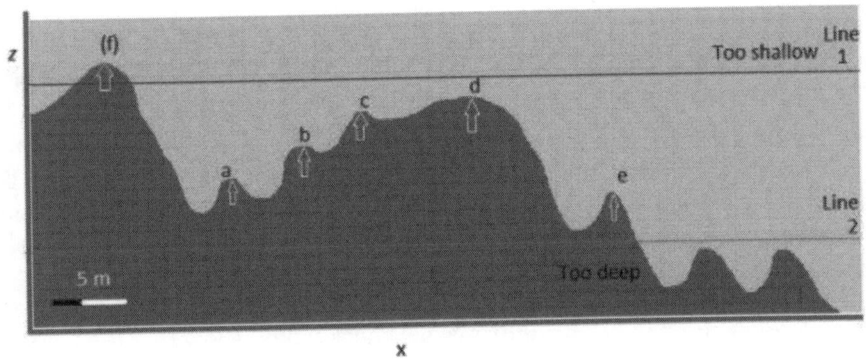

Figure 2. Sample topology of sea bottom with depth limitation lines.

In Fig. 2, line 1 is the 5 m depth mark. Bottom formations above this line (f) are theoretically not included in the data as the surveying vessels cannot sail directly above it. However, due to the side-sweeping technology there would most likely be some data from at least one side of f available. If this is the case then f would also be included in the analysis. The formations below line 2 are below the nominal fareway draft and as such are not included in the analysis at all. Using the method described here, peaks a–e (and f) are analyzed separately.

Step 2: Identify individual peaks from the larger bottom formations

From the larger bottom formations, single peaks from the larger bottom formation (a-e inFig. 2) were cut out as follows:

2a)Selecting the mid-point of the "rock" in the plotted figure (usually the apex) by visual inspection and manual input. Alternatively, an algorithm for peak finding can be used, see Ref. [28].

2b)Defining this point as origo x = y = 0.

2c)selecting all the data points within a radius of r meters from it in the xy-plane for separate analysis.

The apexes had to fulfill the following criteria:

The highest point must be no deeper than fareway nominal depth and the protrusion has to stand out enough from the surrounding features. In practice this means that ~0.5 m "bumps" are included or are included only as a part of a larger bottom formation. Again, an algorithm can be used for large datasets but it has to be tuned properly so that it does not cut the data into too small or large subsets.

Step 3:Define the cut-off method for the individual peaks

In this step the individual peaks are isolated into their own datasets using the following approach:

3a)Selecting all data points less than *r* distance in the xy-plane from each rock midpoint for separate analysis, see Fig. 3. The radius *r* to which select the data around the midpoint or apex (2a) was defined iteratively as follows:

3b)Plotting the selected data separately from the bigger data.

3c)Changing the radius until a visual check would show that the whole of the "rock" would be included and all other apexes would be excluded (except protrusions that are not big enough to be analyzed on their own, see 3a).

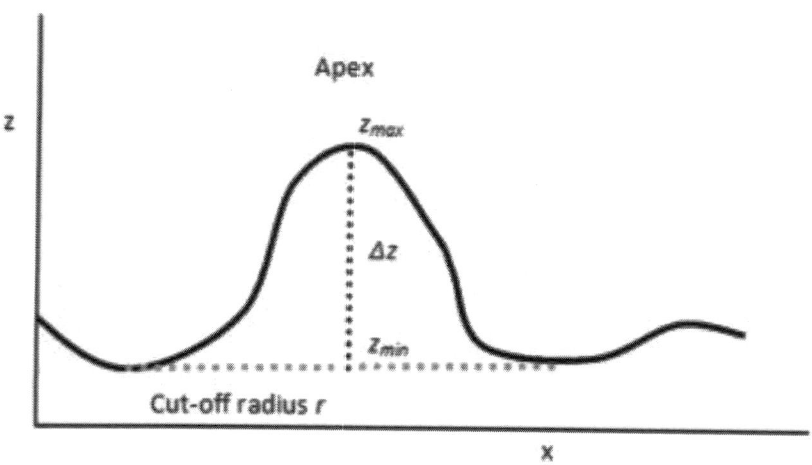

Figure 3. Isolation of individual peaks from larger data.

Alternatively algorithms can be used that e.g. detect changes in gradient. The cut-off formation does not necessarily have to be done as a circle around the apex, working with height/depth curves prominence can be used as a cut-off criteria, see also [28].

Step 4Fit mathematical models to the data and evaluate the models

4a)Define models to be fitted to the data.

4b)Calculate fit.

4c)Calculate goodness of fit, e.g. using coefficient of determination R^2.

4d)Evaluate goodness of fit visually.

4e)Tweak cut-off radius *r* if it improves 4c) and/or 4d) if it does not compromise on 3b).

Evaluating the goodness of the fit (4d) visually is recommended as it is possible to get mathematically well-fitting models that make no practical sense such rock models with the peak kilometers over the sea level. In the following chapter, this four-step methodology is applied to a case-study in the Gulf of Finland.

3. CASE STUDY

The bottom topography data set used in this paper are from hydrographic survey database of Finnish Transport Agency (Liikennevirasto). The data has been collected in controlled hydrographic surveys which fulfill at least Order 1a of IHO's survey standard S-44, see Ref. [12]. Data collection was done from 1999 to 2012 by multibeam (MBES) and multi transducer echo sounder (MTES) systems as fairway surveys. The data consists of discrete xyz-data points of sonar readings, each point corresponding to one sonar signal echo from a bottom shape or object in the water column.

Horizontal resolution of the data sets varies based on survey technique and water depth. Typical sonar footprint ranges between ca. 1 m for MTES to centimeter level for MBES systems in shallow water. The depth value of a sounding is typically detected from the first significant return echo within the footprint. Vertical uncertainty of sounding is typically less than 0.3 m (given a 95% confidence interval), usually much less. Uncertainties regarding relative position and depth are an order of magnitude smaller, and therefore the bottom shape is very well preserved within limitations of horizontal resolution. The data has been post-processed in order to remove obvious outliers above the bottom. Possible outliers close to the bottom have been removed only if independent second source data validates the removal. Since the data is primarily used for nautical charting the depth values do not always represent the most probable bottom depth but shallowest probable bottom depth. Depth difference between these levels is typically few centimeters and does not affect the bottom shape. Parts of the survey data is classified by legislation therefore the true location of individual rocks has been removed but the local shape has been preserved.

4. APPLICATION OF THE FOUR-STEP METHODOLOGY

Steps 1–4 are applied to this data, which is described as follows.

Step 1:Define relevant sea area and find bottom data

The data sets are chosen along two deep draught channels on northern shore of Gulf of Finland that have the highest volume of tanker traffic in Finland:

Sköldvik channel in Porvoo

Design vessel: tanker, L = 300 m, B = 40 m, T = 15.3 m. Maximum authorized draught 15.3 m, safe clearance depth in the outer channel −17.45 m, in the inner channel −16.95 m, minimum width 270 m. Vessel speed 12 knots in the inner channel.

Hamina channel

Design vessel: L = 210 m, B = 32 m, T = 12.0 m. Maximum authorized draught 12.0 m, safe clearance depth in the approach −13.8 m, in the entrance −13.5 m and at the port entrance −13.2 m. Minimum channel width 180 m.

A design vessel indicates the type and the maximum dimensions of vessels for which the fairway is designed.

In Fig. 4 red lines represent the Sköldvik (Porvoo) channel (westernmost) and Hamina (easternmost) channel.

Step 2:

Identify individual peaks from the larger bottom formations

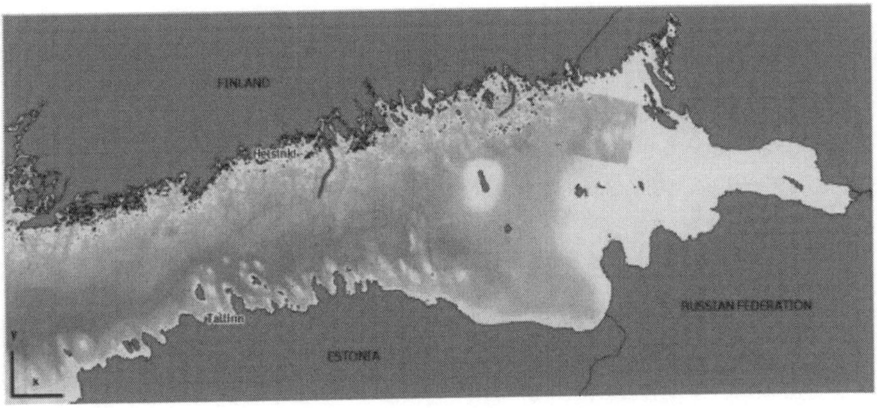

Figure 4. Location of the channels analyzed. Map: © Finnish Transport Agency and Baltic Sea Hydrographic Commission, Baltic Sea Bathymetry Database version 0.9.3.

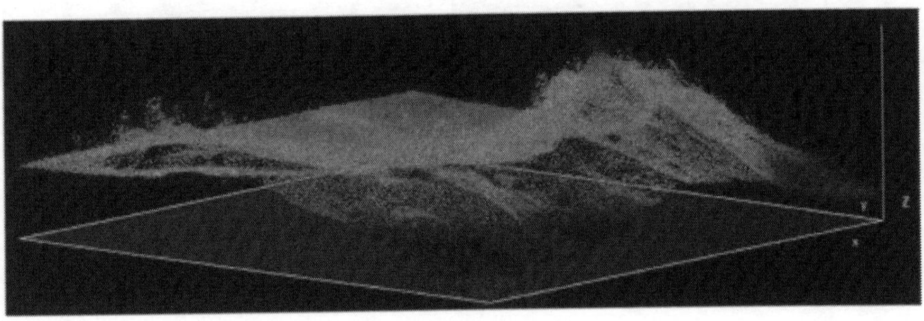

Figure 5. Type 1: Multiple peaks and one clear apex (formation 8).

For the Hamina fareway, the authors were able to isolate 16 different bottom formations with a total of 144 apexes. For Porvoo, the numbers were 13 different bottom formations with a total of 55 apexes. This brings the total number to 199 apexes spread over 29 different formations. The larger formations were labeled with numbers for Hamina and with capital letters for Porvoo. The lower-case letter in parenthesis denotes the individual apexes of the formation in question. A larger formation is typically of the size order of magnitude of 150 m × 150 m × 10 m in the xyz-dimensions. In the data the x-axis values are the easting coordinates and y-axis the northing coordinates for the data points. The z-axis describes the water depth where z = 0 is the sea level. All units are in meters. Looking at the different formations some pattern repeated themselves and the formations in the data were classified into the following three categories:

Type 1: Formations with multiple peaks and one clear apex

Note distorted dimensions: x- and y-axes each cover an area of ~200 m while the z –axis only 6 m. In Fig. 5 the difference between the highest and the second highest peak is >2 m.

Type 2: Formations with multiple roughly even peaks

Type 2 formations may or may not include smaller peaks deeper down such as the following example, which has 11 distinguishable individual apexes, illustrated in Fig. 6.

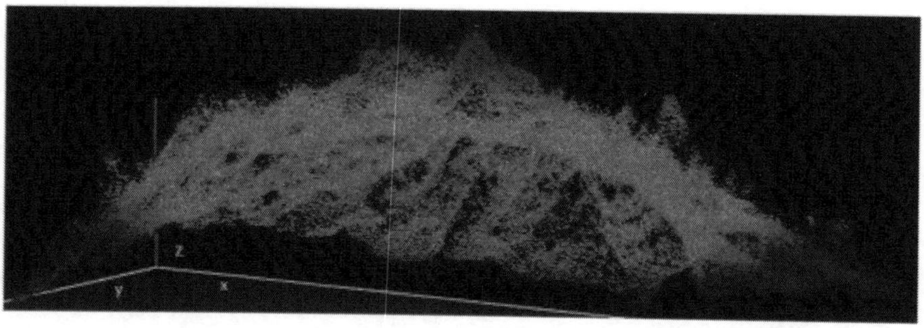

Figure 6. Type 2: Multiple roughly even peaks (formation 4).

Please note that the dimensions of the figure are heavily distorted: The depth-axis (z) runs only from −6 to −14 m ($\Delta z = 8$ m) while the x-axis runs from 0 to 140 m and the y-axis from 0 to 250 m.

Type 3: Solo rounded rock

Besides the multiple peak formations, also some lone rocks were found such as the following example in Fig. 7.

In Fig. 7 the blue dots are the actual data while the surface is a fitted function which is meant to aid in visualization of the form of the rock.

Step 3:Define the cut-off method for the individual peaks

After identifying the peaks in the data, the next step is to define the cut-off radius in the xy-plane for each bottom formation around this apex- or midpoint. The aim is to have each individual bottom shape cut-out into its own data file for separate analysis.

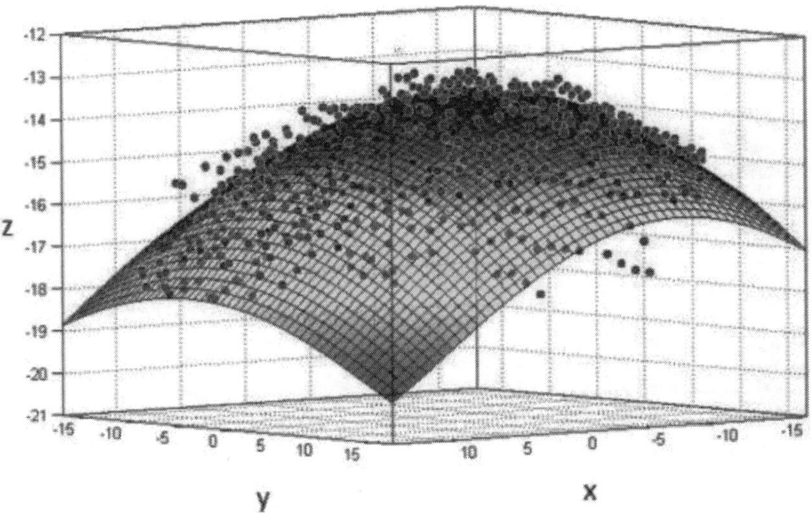

Figure 7. Type 3: Solo round rock (formation V) with data points and a fitted surface.

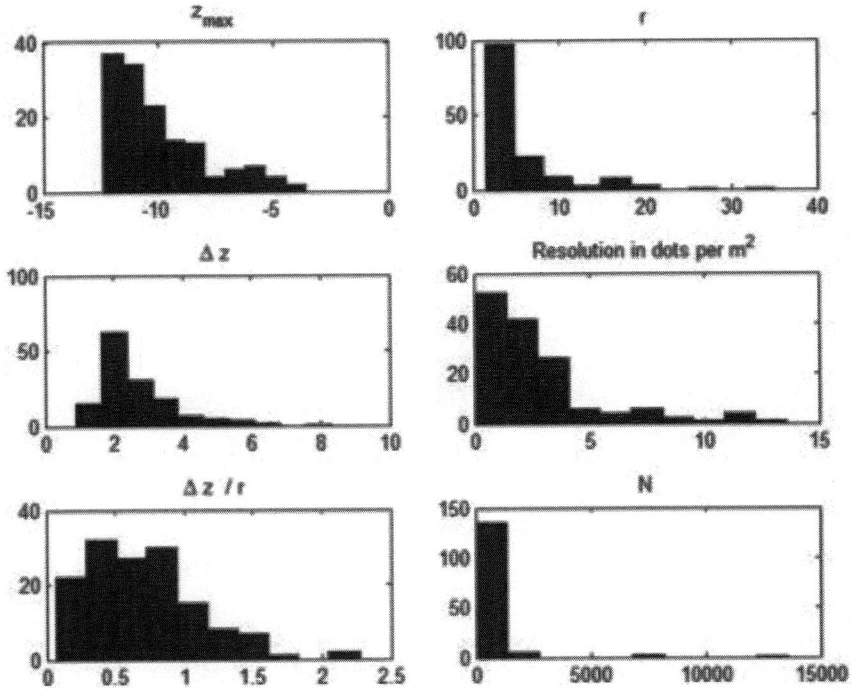

Figure 8. Hamina descriptive statistics.

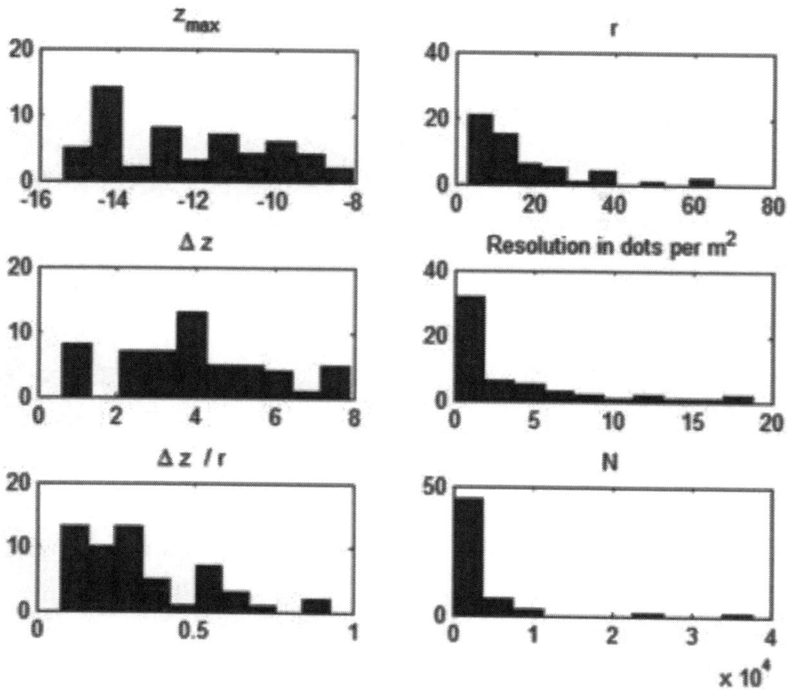

Figure 9. Porvoo descriptive statistics.

In Fig. 8 and Fig. 9 the cut-off radii are presented for the two channels along with other descriptive data. The figures include histograms for the 199 bottom shapes for the following variables:

zmax is the highest point of each peak in meters below the sea level, see Fig. 3.

r is the cut-off radius in meters in the xy-plane around each peak apex or midpoint, seeFig. 3.

Δz is the difference in the highest and deepest point in the z-axis for each bottom formation in meters, see Fig. 3.

Resolution is the number of data points per m^2 in the xy-plane (ie. as seen from directly above).

Δz/r is the depth change divided by the radius of the bottom formation, which tells us how steep the. individual formation is: the lower this values is the blunter the formation and vice versa.

N is the number of the xyz-data points for each individual bottom formation, see the blue dots in Fig. 7 for an example.

The y-axis is the number of observations in each histogram bin.

4.1. Qualitative description of individual peaks

When looking closer at the individual bottom shapes, there are 199 individual peaks which can be classified as follows and are illustrated in Fig. 10:

1.Sharp small peak

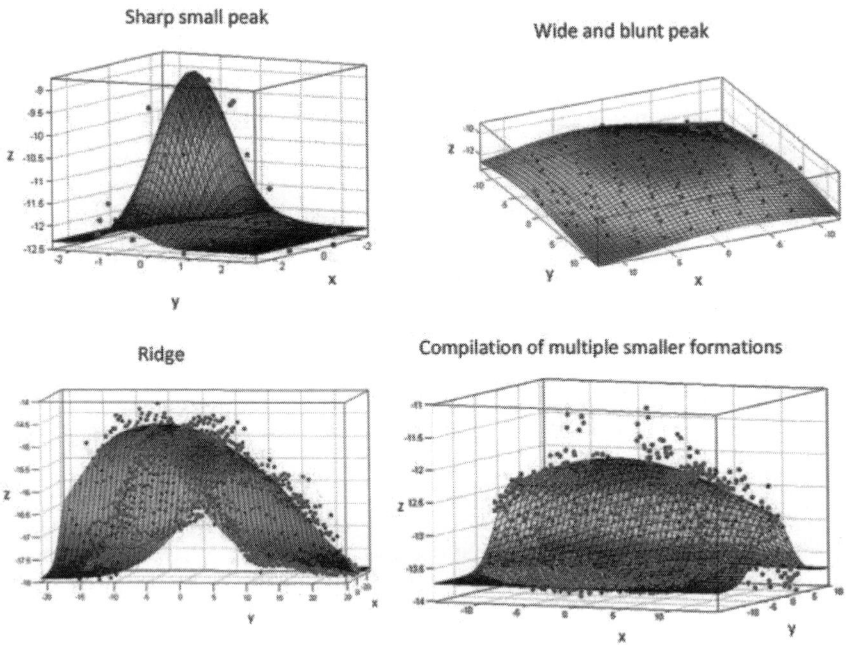

Figure 10. Individual bottom shape types.

These smaller rocks are usually a few meters wide and deep, which means that their size is of the same order of magnitude as the model rocks most commonly used in grounding damage calculations.

2.Wide and blunt peak

These rocks resemble the "shoal" type by Ref. [1]; where the width is around ten times greater than the depth. These rocks may be so wide that they will unlikely cause a spill due to a grounding on a double-hulled vessel, see Ref. [1].

3.Ridge

Ridges are steep to some sides while remaining quite flat in other directions. The models found in the literature review assume symmetry and as such will have problems modeling this type of bottom shapes.

4.Compilation of multiple smaller formations

The smaller tips of the formation could not be modeled as stand-alone bottom shapes: Practical limit is around 1 m protrusion from surroundings or else there is no reasonable fit with the models used here. These formations were analyzed as a cluster of the surrounding shapes, see Fig. 10.

The blue dots are the actual data while the surface is an approximate fit to the data which is meant as a visualization aid. In Fig. 10 the peak types are as follows: Sharp small peak (top left, Formation 10 a), wide and blunt peak (top right, Formation S (a)), ridge (bottom left, Formation D (b)) and compilation of multiple smaller formations (bottom right, Formation 10 (m)). For all of the 199 peaks found in the data, steps 3 and 4 were applied.

Step 4:Fit mathematical models to the data and evaluate the models

To these individual peaks, various functions were fitted according to the approach explained in 4). This data is used to statistically test the current grounding depth and bottom shape assumptions in terms of how well the models describe the data. The data fitting, filtering and testing was done using MATLAB.

The reasoning of the choice of models to fit to the data is as follows: The models should avoid unnecessary complexity so that coefficient interpretation, practical implementation, dissemination of information to other researchers as well as calculations are practically feasible with reasonable effort. As such splines and other more complex models are left as suggestions for future research, see also [31]. The models should also evaluate the current assumptions in literature and preferably at least one model should have a good fit. Models 1 [9] & 3 (cone) are presented in the literature.

Besides these, the authors add two models to the testing: Model 2 (2nd order polynomial) is a more detailed and flexible version of model 1. The scaled and shifted binormal model (4) is selected as it resembles a cone (model 3) with a blunted peak if standard deviations of x and y are small and a 2nd order polynomial equation if the standard deviations are large. Besides this, many things in nature are approximately normally distributed which makes the choice a "natural" one.

The fitted models are as follows:

1.Heinvee et al. [9] model extended in the x-axis direction

$$z = b_0 + b_1(x - x_0)^2 + b_2(y - y_0)^2$$

The origo of the rock is already defined approximately visually in 2a) but as this model is quite sensitive to the exact location of the origo, constants *x0* and *y0* were added so that potentially better fits could be achieved by moving the origo. Besides this another polynomial equation with more terms was tested out to see if adding terms would improve the model fit:

2.2nd order polynomial equation

$$z = b_0 + b_1x + b_2y + b_3x^2 + b_4x\ y + b_5y^2$$

3.Cone

$$z = \frac{(x - x_0)\,h}{r\cos(\arctan((y - y_0)/(x - x_0)))} + h - b_1$$

Where h is the height of the fitted cone and r the radius of the cone at $h = 0$. The constant $b1$ is added to this equation (and equation (4) as well) to counter out that all the observed values of z are negative.

4. Scaled and shifted binormal function

$$z = b_0\frac{1}{2\pi\sigma_x\sigma_y\sqrt{(1-\rho)^2}}\exp\left(-\frac{1}{2(1-\rho^2)}\left(\frac{(x-\mu_x)^2}{\sigma_x^2} + \frac{(y-\mu_y)^2}{\sigma_y^2} - \frac{2\rho(x-\mu_x)(y-\mu_y)}{\sigma_x\sigma_y}\right)\right) - b_1$$

Where $b0$ is the scaling multiplier, $b1$ a constant that counters out that all observed $z$are negative, σ the standard deviation, μ the mean and ρ the correlation between $x$and y.

Note that none of the aforementioned constants in models 1–4 are taken straight from the bottom data. They are the parameters of the best-fitting binormal distribution, which is determined using nonlinear least squares method in MATLAB with the 'fittype' curve- and surface fitting function. This method was applied to all equations (1)–(4). The goodness of the fit was evaluated using the coefficient of determination R^2, which has a best possible value of 1, see Fig. 11 for an example of a fit with high R^2.

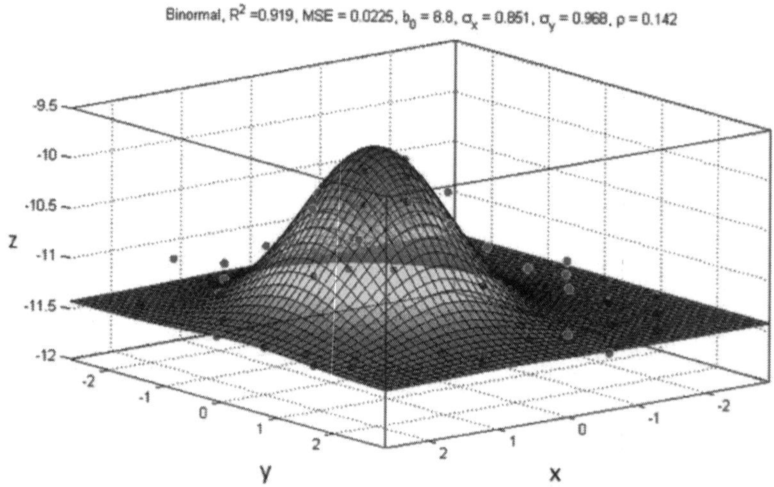

Figure 11. Example of a rock with binormal fit with $R^2 = 0.92$.

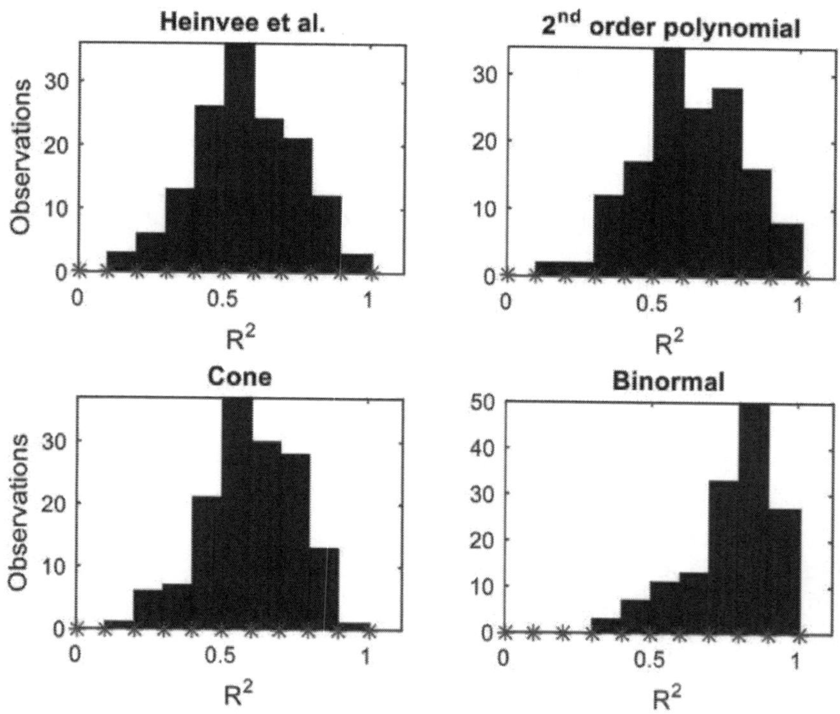

Figure12. R^2 for different models using the Hamina data.

The fitting process yielded the following histograms presented in Fig. 12 for R^2 for the four models that were tested.

Counting the share of all rocks where R^2 exceeded a certain threshold (0.7, 0.8 or 0.9) the following results are obtained, which are summarized in Table 1.

Table 1. Rock model performance for Hamina.

	Heinvee et al.	2nd order polynomial	Cone	Binormal
$R^2 \geq 0.7$	0.25	0.36	0.29	0.76
$R^2 \geq 0.8$	0.10	0.17	0.10	0.53
$R^2 \geq 0.9$	0.02	0.06	0.01	0.19

Generally the binormal function is superior compared to the other three models in terms of R^2: 76% of all modeled rock formations had $R^2 \geq 0.7$, 53% $R^2 \geq 0.8$ and in 19% of cases $R^2 \geq 0.9$. The other three models give smaller shares for the thresholds. In 95.14% of the cases the binormal function had the best R^2 of all 4 models when comparing each case separately side-by-side.

In Fig. 13 the upper leftmost corner shows the radius plotted against R^2 of the binormal function. Rocks with radii greater than 10 m consistently have an R^2 of at least 0.8 while smaller rocks have R^2 of everything from less than 0.4 up to 1. There seems to be no clear connection between R^2 and the resolution (upper right corner in Fig. 13), individual rock depth change (right middle in Fig. 13) nor the apex minimum depth (lower right corner in Fig. 13). See Appendix 1 for histograms of the parameters of the binormal functions for all analyzed rocks. For Porvoo the results are illustrated in Fig. 14 and Table 2.

Table 2. Rock model performance for Porvoo.

	Heinvee et al.	2nd order polynomial	Cone	Binormal
$R^2 \geq 0.7$	0.53	0.76	0.36	0.84
$R^2 \geq 0.8$	0.27	0.56	0.13	0.65
$R^2 \geq 0.9$	0.05	0.16	0.02	0.20

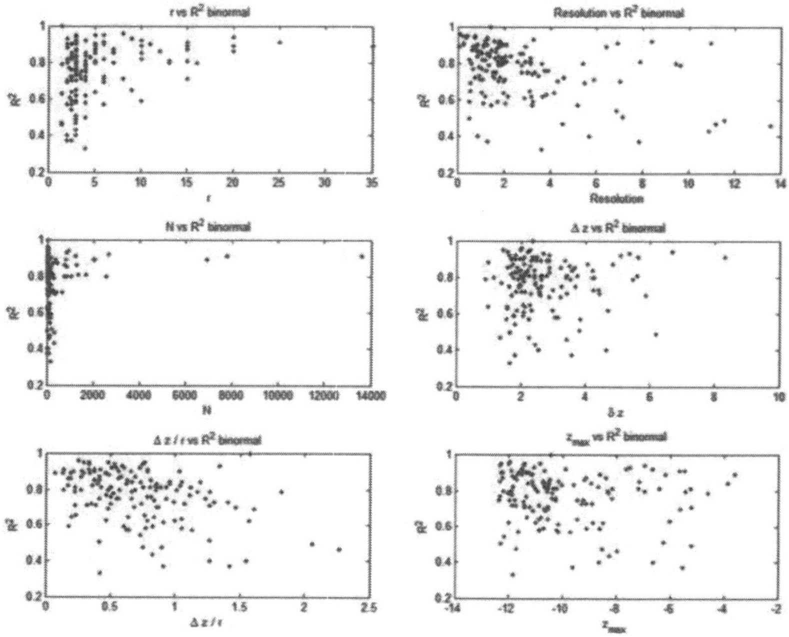

Figure 13. Descriptive statistics of Hamina individual peaks.

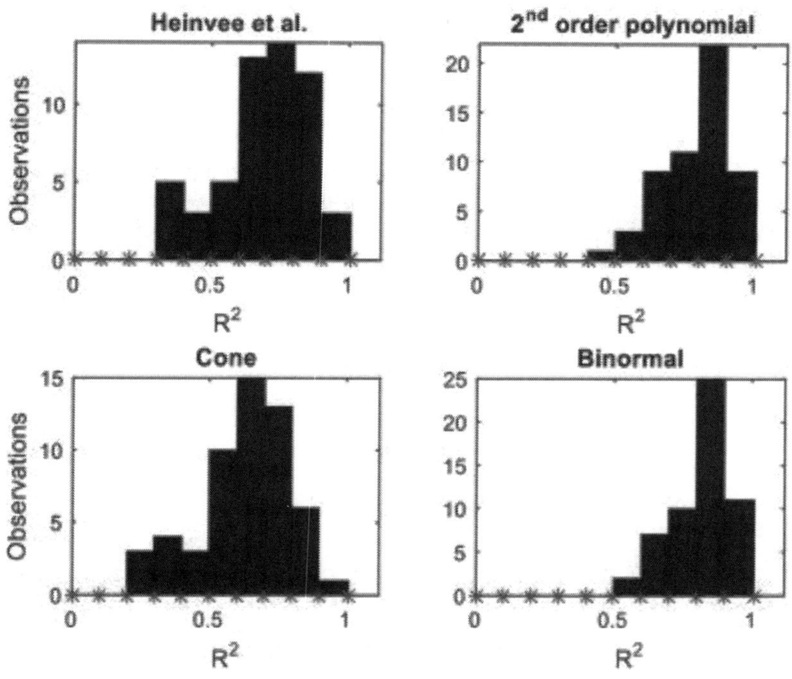

Figure 14. R^2 for different models using Porvoo data.

For Porvoo the binormal model gives superior results in terms of the share of all cases where R^2 exceed a certain threshold: 84% of cases have $R^2 \geq 0.7$, in 65% of cases $R^2 \geq 0.8$ and in 20% $R^2 \geq 0.9$. All other models have a smaller share of cases where these thresholds were exceeded.

The binormal model was the best in terms of R^2 in 71% of all cases when comparing the four models to one another case-by-case. The threshold for getting consistently R^2 of at least 0.8 for is $r > 17$ m, see Fig. 15.

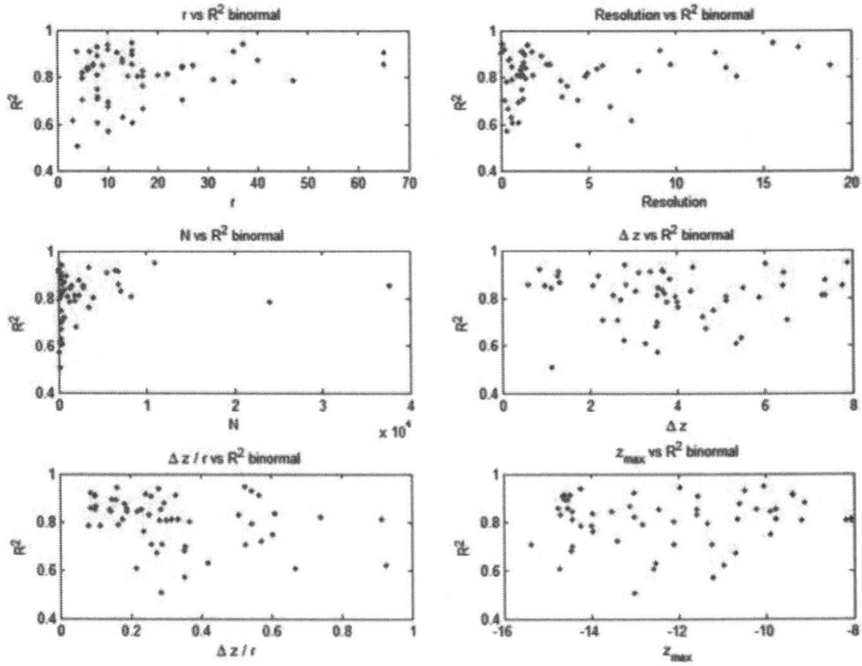

Figure. 15. Descriptive statistics of Porvoo individual peaks.

In Fig. 15 the resolution vs R^2 shows a slightly increasing trend: Peak datasets with resolution over 8 observations per m^2 have consistently a $R^2 \geq 0.8$ while the lower resolution ones have R^2 values ranging from <0.6 to ~1.

4.2. Grounding depth based on z_{max}

In order to test whether the grounding depth follows a triangular or a uniform distribution[30] and [13], the following null- and alternative hypothesis were formulated and tested using the Kolmogorov–Smirnoff goodness-of-fit test:

H_0 = *zmax*follows a specified distribution (triangular or uniform)

H_1 = *zmax*does not follow the specified distribution

The assumptions for the triangular distribution are as follows: Lower bound and most likely value is the smallest value of *zmax*, the upper bound and least likely value is the largest observed value of *zmax*. This means that it more likely to ground on deeper rocks than on rocks with peaks in shallow waters, see Fig. 8 and Fig. 9 for histograms of *zmax*. The following results were obtained when constructing cumulative frequency distributions for the theoretical triangular distribution and the data:

Looking at Fig. 16, the biggest discrepancies between the observed Hamina data (+) and the theoretical triangular distribution (solid line) is at around −10.5 m while the uniform distribution shows the largest difference in the −11 to −8 m range. Porvoo shows the largest difference between the data and the distributions at around −15 m to −14 m. For Hamina, the p-value of the test is 0.0048 for the triangular and $3.5 \cdot 10^{-10}$ for the uniform distribution, meaning that the null hypothesis that grounding depth follows a triangular distribution is rejected at a significance level of 0.05.

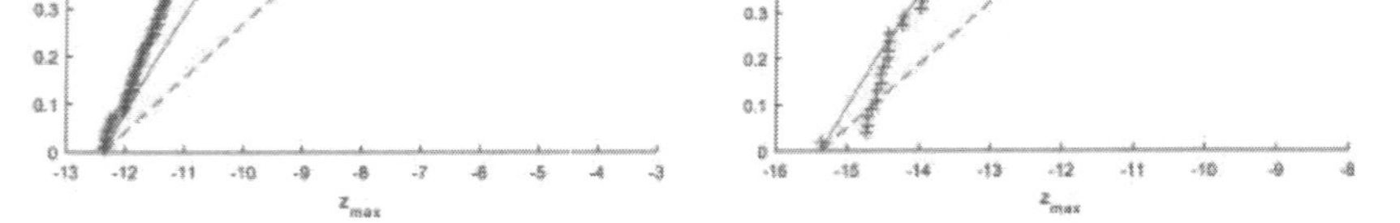

Figure. 16. Cumulative distribution functions for *zmax* Hamina (left) and Porvoo (right).

For Porvoo, the p-value of the test is 0.293 for the triangular and 0.420 for the uniform distribution, meaning that the null hypothesis that grounding depth follows any of these two distributions is not rejected at a significance level of 0.05. Combining the data from Porvoo and Hamina the following results are obtained:

Looking at Fig. 17 the largest discrepancy between the theoretical triangular distribution and the observed values is at around −12 m mark while the data overlaps quite well with the triangular distribution from −10 to −2 m. The uniform distribution deviates most from the data in the −10 m to −8 m depth range. The p-value of the test is $3.19 \cdot 10^{-8}$ for the triangular and $1.02 \cdot 10^{-6}$ for the uniform distribution meaning that the null hypothesis that the combined dataset grounding depth follows any of these two distributions is rejected at a significance level of 0.05.

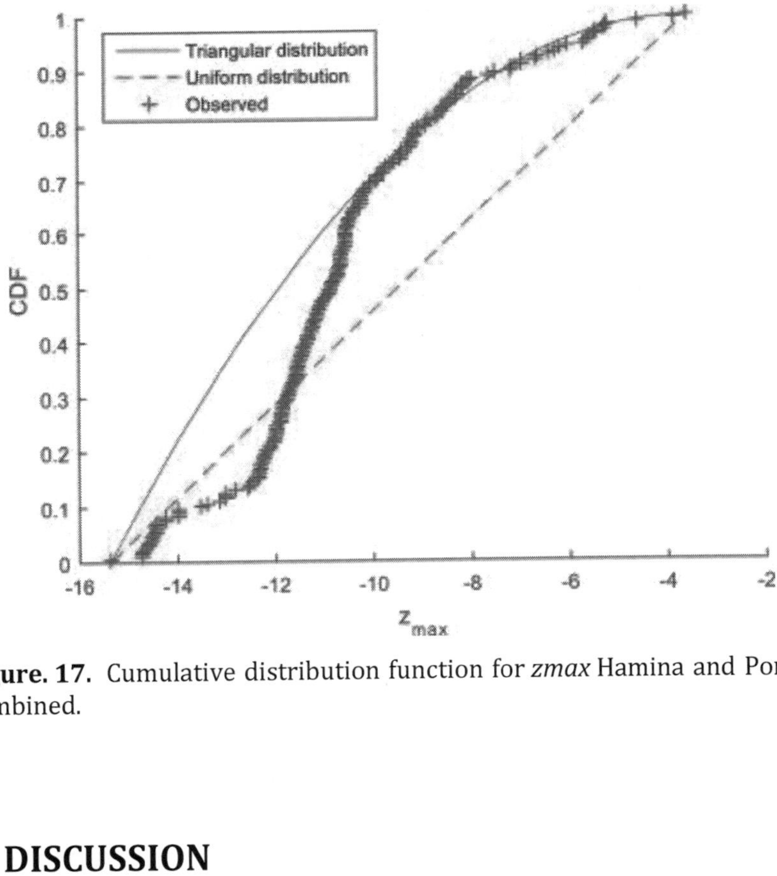

Figure. 17. Cumulative distribution function for *zmax* Hamina and Porvoo combined.

5. DISCUSSION

The four-step approach to mathematically model bottom shapes was presented and applied to two case studies from Finnish waters. Sea bottom varies in shape and size, range from sharp rocks a few meters wide to shallow round shapes with a radius of several tens of meters; thus in the data examples that size-wise roughly correspond to the conical rock models, the different rock sizes A-D of [9] as well as the three bottom shape cases presented in Ref. [1] were found. In addition, asymmetrical shapes such as ridges and groups of smaller formations right next to one another were observed. The data shows a difference in the size of bottom shapes between Porvoo and Hamina: the bottom shapes in Porvoo tended to be much wider (bigger *r*) and not as sharp (smaller $\Delta z/r$) as the ones in Hamina. As such there is a need to take local conditions into account when performing grounding damage analysis. This shows the need for more rock models to supplement the commonly assumed single cone shapes and symmetrical polynomial equations. An example of a model that fulfills the aforementioned criteria is the binormal function. Note that some of the cones presented in literature (such as [21] are qualitatively assessed very similar to a binormal function; However these cones are symmetrical in the *xy*-plane whereas the binormal function is more flexible in terms of modeling the actual bottom topology. The data shows that this flexibility is needed. A pure, unblunted cone does not fit to the data well in terms of R^2 in most cases: it achieved a fit of 0.9 or better in only 1–2% of cases and fit of at least 0.8 in only 10–13% of cases. The results polynomial model fits were mixed: Relatively good fit for Porvoo (over half $R^2 \geq 0.8$) while for Hamina $R^2 \geq 0.8$ only in 17% of cases. The [9] model was on average slightly worse than the 2nd order polynomial model with more terms. The binormal function gives better results and in comparison has the best fit in the vast majority of cases. For larger rocks (10 m for Hamina, 17 m for Porvoo) fits of $R^2 \geq 0.8$ are consistently obtained, whereas the small rocks show R^2 ranging from 0.3 to 1. The binormal function a flexible and a good alternative to the conical model and/or polynomial models. The binormal function can generally be used to model the cases where other models have a good fit, see Fig. 18 and Fig. 19.

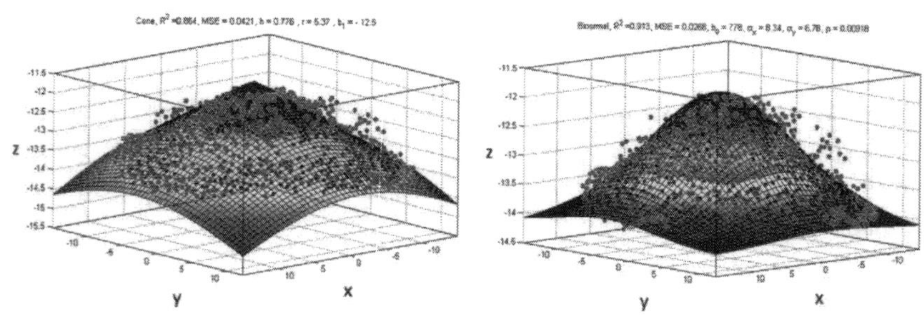

Figure. 18. Formation 16 with fitted cone (left) and binormal (right) models.

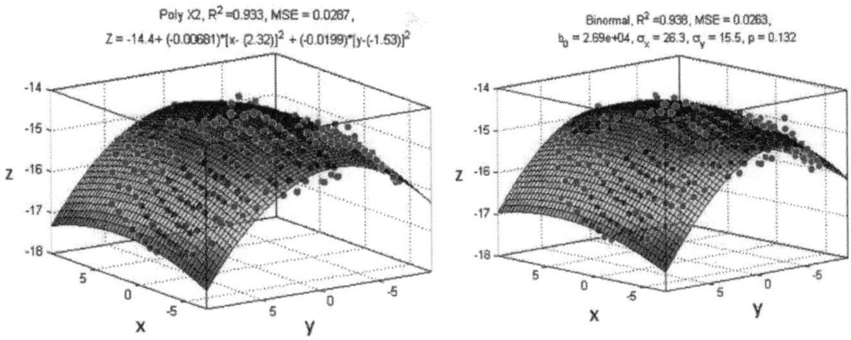

Figure. 19. Formation D (a) with fitted 2nd order polynomial (left) and binormal (right) models.

As Fig. 18 shows, the cone model gives $R^2 = 0.864$ and the binormal model $R^2 = 0.913$ respectively. Similarly Fig. 19 shows that the 2nd order polynomial model gives $R^2 = 0.933$ and the binormal model $R^2 = 0.938$. Thus, binormal model outperformes slightly the cone and polynomial models. The tested cone type here is a pure, unblunted cone. The R^2 could most likely be improved by blunting or otherwise modifying the tip part of the cone. However the fit results are not encouraging; if cones in general does correspond with the data, one would expect higher R^2 values than the current results. Any asymmetry in the bottom shape will still result in a bad fit for the cone model and would require the model to be changed to elliptical instead. Blunting the cone heavily such as in Ref. [21] brings the cone very close – if not identical – to a binormal function. The authors recommend using the shifted and scaled binormal function in modeling bottom shapes that can be assumed to be similar to the ones presented in this paper. See Appendix 2 for examples of a small, sharp rock and a wide, blunt rock and their parameters.

Note that the coast of Finland has been strongly shaped by the latest ice age and as such the analysis presented as steps 1–4 should be repeated in other sea areas for more general recommendations.

The incapability of the binormal model to model all of the smaller rocks leads us to suggest investigating the fit of other more complex or non-parametrical models such as the approach presented in Ref. [31]. The selected data here does not cover stranding, which is also left as a suggestion for the future research.

The R^2 does not tell the whole truth: A good fit in statistical terms does not warrant a good representation of reality. For example Hamina Formation 2 m has $R^2 = 1$ but the height of the fitted binormal function is several hundred meters. The opposite is also true: what looks qualitatively to be a good fit might be ruined by "outliers" at the base of the rock, yielding a low R^2 score. As R^2 does not straightforward tell the exact goodness of the fit in applied risk analysis. The real question is whether using these fitted models would result in the same damage as a raw data rock when analyzing ship grounding damage with models such as the one by Ref. [9]. In other words, the authors highly recommend an investigation on the resulting damage between realistic and idealized ground shapes. The numerical methods to estimate the grounding damage are getting mature as the understanding of the failure process of material has increased and computational power increases as well. Thus, the data from this paper should be applied to numerical simulations to show the effects of mathematical idealizations on ground shape on the simulated consequences. This would increase the reliability of the simulations. However, this is left for future work.

Regarding the grounding depth, the Kolmogorov–Smirnoff test results support both the triangular as well as the uniform distribution assumption for Porvoo data: This means that according to the hypothesis tests any of these two distributions can be used to describe grounding depth in Porvoo. For Hamina and the combined dataset the conclusion is opposite as both distributions were rejected in the hypothesis testing. Overall analyzing more data is suggested as future research.

6. CONCLUSIONS

The paper presented a methodology to measure the actual bottom topology of specific ship routes. The methodology was used at two locations in Baltic Sea on the coastal line of Finland. The measured shape was compared with simplified mathematical shapes from literature, i.e. cone and parabolic. The curve fitting show that these simple forms do not fit the measured data very accurately in most cases. This should cause some uncertainty to the input to the Finite Element simulations of the actual grounding damage. Alternative model is presented based on a scaled and shifted binormal function and it is shown that it outperforms the other available models in terms of statistical goodness-of-fit. Regarding the grounding depth, the Kolmogorov–Smirnoff test results support the triangular assumption for Porvoo data but not for Hamina data, leaving the question ambiguous. Analyzing more data is suggested as future research.

This data should be used in conjunction with the Finite Element simulations to clarify how much the absorbed energy is under- or overestimated using these simplified models. This is left for future work. Furthermore, the authors propose the following for future research including: Investigating for different ships how probable it is to hit each the different bottom shapes. The potential effects of grounding angle should be investigated as the data shows asymmetrical bottom shapes. This asymmetry might be bound to compass heading due to local geological factors; e.g. in Finland glaciers retreated from South-West towards North-East during the last ice age, causing directed erosion. Groundings and strandings on slopes leading up to fast land and non-solid bottom shapes could also be studied. Furthermore the authors propose investigating what spatial correlation and patterns exists between individual peaks in a larger formation, see e.g. Ref. [28]. This is also left for future work.

ACKNOWLEDGMENTS

This paper has received funding from Kotka Maritime Research Center Merikotka. The authors would like to thank Finnish Transport Agency for the data and the friendly spirit of co-operation which made this study possible. For programming and analyzing tips the authors would like to thank Johannes Haataja, Jaripekka Juhala and Daniel Laajala. For feedback on writing and how to frame the study the authors would like to thank Otto Puolakka, Heikki Remes, Teemu Manderbacka and Markus Ahola.

REFERENCES

1. Alsos H, Amdahl J. On the resistance of tanker bottom structures during stranding. Mar Struct 2007;20:218e37.
2. Cerup-Simonsen B. Ship grounding on rockdI. Theory. Mar Struct 1997a;10(7):519e62. [3]
3. Cerup-Simonsen B. Ship grounding on rockdII. Validation and application. Mar Struct 1997b;10(7):563e84.
4. Cerup-Simonsen B, Tornqvist R, Lützen M. A simpli € fied grounding damage prediction method and its application in modern damage stability requirements. Mar Struct 2009;22(1):62e83.
5. Choung J, Nam J-M, Ha T-B. Assessment of residual ultimate strength of an asymmetrically damaged tanker considering rotational and translational shifts of neutral axis plane. Mar Struct 2012;25(1):71e84.

6. Cox M, Budhu M. A practical approach to grain shape quantification. Eng Geol 2008;96(1e2):1e16.

7. Cruden D. The shapes of cold, high mountains in sedimentary rocks. Geomorphology 2003;55(1e4):249e61.

8. Heinvee M, Tabri K. A simplified method to predict grounding damage of double bottom tankers. Mar Struct 2015;43: 22e43.

9. Heinvee M, Tabri K, Korgesaar M. A simpli ~ fied approach to predict the bottom damage in tanker grounding. In: Proceedings of the 6th International Conference on Collision and Grounding of Ship and Offshore Structures, Trondheim 17e19; June 2013. Available online: http://www.ntnu.no/iccgs [accessed 23.01.14].

10. Hu Z, Amdahl J, Hong L. Verification of a simplified analytical method for predictions of ship groundings over large contact surfaces by numerical simulations. Mar Struct 2011;24(4):436e58.

11. Hong L, Amdahl J. Crushing resistance of web girders in ship collision and grounding. Mar Struct 2008;21(4):374e401.

12. International Hydrographic Organization (IHO). IHO standards for hydrographic surveys. Special Publication no 44. 5th ed. International Hydrographic Bureau; 2008. p. 15e6. Available online: http://www.iho.int/iho_pubs/standard/S-44_5E.pdf [accessed 20.05.15].

13. Jalonen R. Riskianalyysi ROPAX-aluksen karilleajotapahtumassa. Licenciate Thesis. Aalto University; 2003.

14. Kitamura O. FEM approach to the simulation of collision and grounding damage. Mar Struct 2002;15:403e28.

15. Klanac A, Jalonen R, Varsta P. Multi-stakeholder decision-making in the risk-based design of a RO-PAX double bottom for grounding. Proc IMechE Part M J Eng Mar Environ 2006;221:1e15.

16. Lehmann E, Peschmann J. Energy absorption by the steel structure of ships in the event of collisions. Mar Struct 2002; 15(4e5):429e41.

17. Liu B, Soares CG. Simplified analytical method for evaluating web girder crushing during ship collision and grounding. Mar Struct 2015;42:71e94.

18. Marinatos JN, Samuelides MS. Towards a unified methodology for the simulation of rupture in collision and grounding of ships. Mar Struct 2015;42:1e32.

19. Młynarczuk M. Description and classification of rock surfaces by means of laser profilometry and mathematical morphology. Int J Rock Mech Min Sci 2010;47(1):138e49.

20. Moskalik M, Błaszczyk M, Jania J. Statistical analysis of Brepollen bathymetry as a key to determine average depths on a glacier foreland. Geomorphology 2014;206:262e70.

21. Naar H, Kujala P, Cerup-Simonsen B, Ludolphy H. Comparison of the crashworthiness of various bottom and side structures. Mar Struct 2002;15(4e5):443e60.

22. Nguyen T-H, Amdahl J, Leira BJ, Garre L. Understanding ship-grounding events. Mar Struct 2011;24(4):551 e69.

23. Otto S, Pedersen PT, Samuelides M, Sames PC. Elements of risk analysis for collision and grounding of a RoRo passenger ferry. Mar Struct 2002;15(4e5):461e74.

24. Pedersen PT. Review and application of ship collision and grounding analysis procedures. Mar Struct 2010;23(3):241e62.

25. Prasicek G, Otto J-C, Montgomery D, Schrott L. Multi-scale curvature for automated identification of glaciated mountain landscapes. Geomorphology 2014;209:53e65.

26. Rawson C, Crake K, Brown A. Assessing the environmental performance of tankers in accidental grounding and collision. SNAME Trans 1998;106:41e58. [

27. Robin P-Y, Charles C. Quantifying the three-dimensional shapes of spheroidal objects in rocks imaged by tomography. J Struct Geol 2015;77:1e10.

28. Sadahiro Y, Masui M. Analysis of qualitative Similarity between surfaces. Geogr Anal 2004;36:217e33.

29. Smith J. Determining the size and shape of blocks from linear sampling for geotechnical rock mass classification and assessment. J Struct Geol 2004;26(6e7):1317e39.

30. Tikka K, Fischbeck P, Bergquist J, Tuckel E. Analysis of hypothetical groundings in US waters. Mar Struct 2002;15(4e5): 499e513.

31. Torppa J, Valkonen J, Muinonen K. Three-dimensional stochastic shape modelling for potato tubers. Potato Res 2006;49: 109e18.

32. Trier Ø, Zortea M, Tonning C. Automatic detection of mound structures in airborne laser scanning data. J Archaeol Sci Rep 2015;2:69e79.

33. van de Wiel G, van Dorp J. An oil outflow model for tanker collisions and groundings. Ann Oper Res 2009;26(1):1e26.

34. Wang G, Arita K, Liu D. Behavior of a double hull in a variety of stranding or collision scenarios. Mar Struct 2000;13(3): 147e87.

35. Wang G, Chen Y, Zhang Z, Peng H. Longitudinal strength of ships with accidental damages. Mar Struct 2002;15(2):119e38.

36. Yu Z, Hu Z, Wang G. Plastic mechanism analysis of structural performances for stiffeners on bottom longitudinal web girders during a shoal grounding accident. Mar Struct 2015;40:134e58.

37. Yu Z, Hu Z, Amdahl J, Liu Y. Investigation on structural performance predictions of double-bottom tankers during shoal grounding accidents. Mar Struct 2013;33:188e213.

38. Zhang Q, Wu A, Zhang L. Statistical analysis of stochastic blocks and its application to rock support. Tunn Undergr Space Technol 2014;43:426e39.

39. Zhang S. Plate tearing and bottom damage in ship grounding. Mar Struct 2002;15(2):101e17.

40. Zhu L, James P, Zhang S. Statistics and damage assessment of ship grounding. Mar Struct 2002;15(4e5):515e30.

6

CHAPTER

The Stiffness and Damping Characteristics of a Dual-Chamber Air Spring Device Applied to Motion Suppression of Marine Structures

Xiaohui Zeng [1,*], Liang Zhang [1], Yang Yu [2], Min Shi [1] and Jifu Zhou [1]

[1]*Institute of Mechanics, Chinese Academy of Sciences, Beijing 100190, China*

[2]*Collaborative Innovation Center for Advanced Ship and Deep-Sea Exploration, State Key Laboratory of Hydraulic Engineering Simulation and Safety, Tianjin University, Tianjin 300072, China*

ABSTRACT

Dual-chamber air springs are used as a key component for vibration isolation in some industrial applications. The working principle of the dual-chamber air spring device as applied to motion suppression of marine structures is similar to that of the traditional air spring, but they differ in their specific characteristics. The stiffness and damping of the dual-chamber air spring device determine the extent of motion suppression. In this article, we investigate the stiffness and damping characteristics of a dual-chamber air spring device applied to marine structure motion suppression using orthogonal analysis and an experimental method. We measure the effects of volume ratio, orifice ratio, excitation amplitude, and frequency on the stiffness and damping of the dual-chamber vibration absorber. Based on the experimental results, a higher-order non-linear regression method is obtained. We achieve a rapid calculation model for dual-chamber air spring stiffness and damping, which can provide guidance to project design.

Keywords: stiffness and damping; orthogonal analysis; vibration suppression; floating marine structure; dual-chamber air spring

1. INTRODUCTION

Marine structure safety, especially in the deep sea, is an important concern. The deep-sea floating platform is one of the most important large structures used in ocean energy exploitation. Dynamic responses due to the wind and wave effects must be considered to ensure a safe design [1]. The combined action of complex multiple loads could cause a large vibration amplitude in marine structures [2,3]. Large amplitude and alternating vibration are the main detrimental effects that reduce the safety and fatigue life of a deep-sea platform. In the field of civil engineering, engineers use a wide variety of energy-absorbing devices to reduce the vibration of building structures. These energy-absorbing devices may also be applied to the problem of vibration suppression of deep-sea marine structures.

The vibration of a floating platform could be reduced using active or passive controls [4,5] and various devices have been proposed. A Tuned Liquid Column Damper (TLCD), a kind of passive vibration control device to suppress movement, was proposed by Sakai [6] in 1989. Lee *et al.* [7,8] first used the TLCD to suppress the surge and sway motions vibration of a Tension Leg Platform (TLP). Taflanidis *et al.* [9] further developed a simulation-based method for the design of mass dampers applied for the response mitigation of tension lag platforms. Lee *et al.* [10] experimentally studied the harmonic responses of the TLCD for wind excitations. Tanmoy *et al.*[11] studied the effective performance of TLCD and another passive vibration-mitigating device, a tuned liquid column ball damper for the control of wave-induced vibration. Zeng *et al.* [12] invented a new type of energy-absorbing device, the S-shaped TLCD. The S-shaped TLCD could effectively suppress the horizontal movement and vertical in-plane rotation of a TLP. Although these devices were effective for the suppression of surge and sway motion, TLCD was ineffective at the suppression of vertical movement, and sometime even enhanced this movement [10]. The vertical movement of the floating structures significantly impacts the strength and fatigue of the tension leg and mooring system, and the inability of these devices to suppress this movement could decrease the safety and service life of the structure.

Vibration-induced vertical motion of TLPs could be balanced by using absorbers with air springs. Dual-chamber air springs are an effective energy-absorbing device used as a key component for vibration isolation. Rijken, Bian, and Spillane *et al.* [13,14,15] applied a system of vibration absorbers using dual-chamber air springs and water columns to suppress resonant motions and studied the effects of the orifice ratio for structural damping. Bachrach and Rivin [16] studied the complex dynamic stiffness of the damper spring, a function of the excitation frequency. The experimental results measured by Kim and Lee [17] for a dual-chamber pneumatic spring exhibited significant amplitude-dependent nonlinear behavior. Jing *et al.* [18,19,20] proposed a characteristic output spectrum nonlinear (nCOS) method. The nCOS method is based on the theory of Volterra series expansion. They used the nCOS method for analysis and design of an air spring in a nonlinear vehicle suspension system. These studies focused on traditional forms of a dual-chamber air spring.

A dual-chamber air spring device applied to marine structure motion suppression requires a gas-liquid coupling air spring structure. This differs from the traditional air spring both in structure type and in the vibration characteristics. The compression of the gas in the traditional air spring was controlled by a piston, which acted on the pressure surface of the air spring [21,22,23]. As shown in Figure 1a, the displacement of the piston, x_0, causes a compression of the air in the upper chamber. The pressure variation of the air spring chamber is of the same phase and amplitude as the external load. For the gas-liquid coupling air spring device applied to motion suppression of marine structures, a liquid column can also move in an independent manner. As shown in Figure 1b, the amount of compressed air in the air chamber is the relative displacement between the liquid column and the top of the air spring, *i.e.* $x_1 - x_0$. The movement of the liquid column (similar to the piston) and the external excitation can differ in phases and amplitude. The stiffness and damping characteristics of this kind of dual-chamber air spring have not been investigated.

The two main determinants of structure movement suppression effect are the stiffness and damping of the energy-absorbing device. Previous studies of damping and stiffness focused on the traditional air spring, and used an approximation to simulate air flow through the orifice plates [24,25,26]. These results may not apply to the gas-liquid coupling dual-chamber air spring device. Moreover, few factors were investigated and analyzed. There is not a comprehensive study on which factors affect the stiffness and damping characteristics of dual-chamber air spring. Further studies must be carried out on the gas-liquid coupling dual-chamber air spring device.

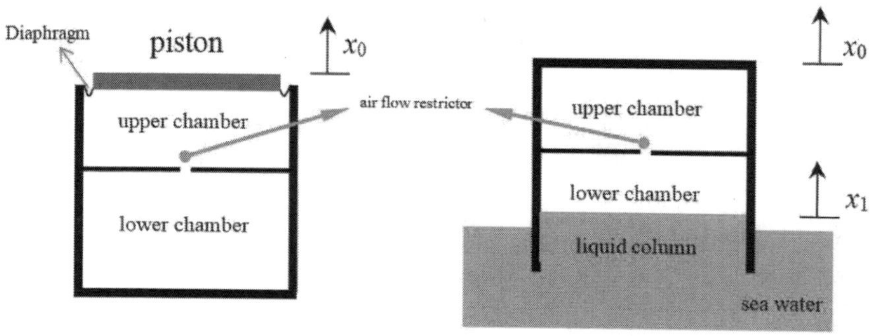

Figure 1. Schematic diagram of two kinds of dual-chamber air spring: (**a**) the traditional device; (**b**) a gas-liquid coupling device.

Here, we investigate the damping and stiffness of a dual-chamber air spring device applied to motion suppression of marine structures. We used orthogonal analysis and an experimental method and found that stiffness and damping are a function of the volume ratio, orifice ratio, excitation amplitude, and frequency. We then measured the effects of these four factors on dual-chamber vibration absorber stiffness and damping. Based on the experimental results, we used higher-order non-linear regression method and generated a rapid calculation model for dual-chamber air spring stiffness and damping, which can provide guidance for future project design.

2. THEORETICAL ANALYSIS

The gas-liquid coupling, dual-chamber, air spring energy-absorbing device has two chambers. The upper chamber functions for air storage and has a fixed volume, while the volume of the lower chamber, in contact with the water, varies with the motion of the water column. The gas and the oscillating liquid column are coupled and this device uses the interaction of oscillations of the liquid column and gas to achieve floating structure motion suppression.

The performance and design of the dual-chamber air spring device are usually modeled after a spring damper system. The characteristics of the damping device are described by the spring stiffness and damping coefficient [26,27,28,29,30]. In this work, the dual-chamber air spring device is a complex fluid-gas coupling system, but its dynamic behavior can also be characterized by the spring stiffness and damping coefficient. The physical model of the common spring damping system can be described using a parallel model or as a series model. Here, we explore the advantages and disadvantages of the two models to characterize the dynamic characteristics of the dual-chamber air spring device applied to motion suppression of marine structures.

In this section, three kinds of energy dissipation expression are derived: a series model, a parallel model, and work done by excitation force.

2.1. Parallel Model

As shown in Figure 2, the standard equation of the model is:

$$M\ddot{X} + C\dot{X} + KX = F \tag{1}$$

where M represents the overall mass of the vibration absorber structure, C represents the overall damping of the structure, and Krefers to the overall stiffness of the structure.

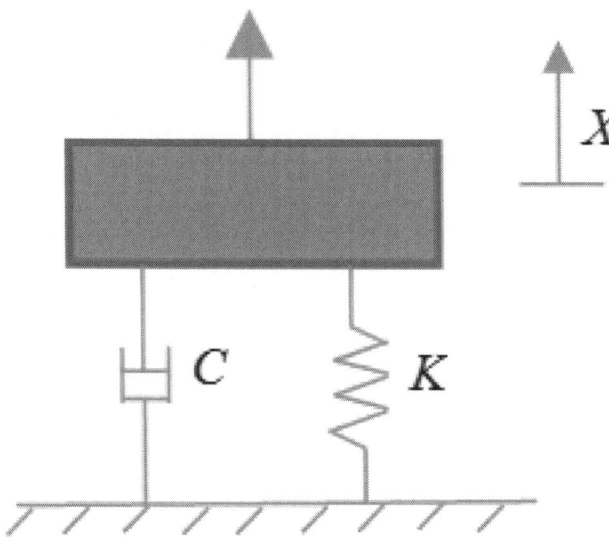

Figure 2. Parallel model.

The structure stiffness in the above equation changes with the air pressure in the vibration absorber air chamber. The change in air pressure is caused by the relative motion of the liquid column in the air chamber. The damping has two components. Some damping is due to fluid vortex shedding at the bottom of the vibration absorber and the friction between the outer wall and water and this damping is related to the motion of the vibration absorber structure. Damping is also caused by the effect of the air in the air chamber and the orifice as well as the friction between inner wall and water, and this damping is related to the motion of the liquid column in the vibration absorber to the vibration absorber.

For harmonic excitation F=F0eiωt, the structural response X is also a harmonic motion which can be represented as X=X0ei(ωt−φ), substituted into Equation (1) and using Euler's Formula, eiφ=cosφ+isinφ, the stiffness and damping of the vibration system can be written as:

$$K = \frac{F_0}{X_0}\cos\varphi + \omega^2 M \tag{2}$$

$$C = \frac{1}{\omega}\frac{F_0}{X_0}\sin\varphi \tag{3}$$

The mass matrix can be M=[m1,m2,m3,m4,m5,m6,m7]T. In the experiment, the mass of each different air chamber length structure is shown as follows:

$m_1 = 5.769, m_2 = 5.967, m_3 = 6.178, m_4 = 6.386, m_5 = 6.593, m_6 = 6.773, m_7 = 7.001$, in kg.

The periodical energy consumption of damping is:

$$Q = \int C\dot{X}dX = \int_0^T \left(C\dot{X}\right)\dot{X}dt = \int_0^T C\dot{X}^2 dt = C\int_0^T \dot{X}^2 dt = \frac{1}{\omega}\frac{F_0}{X_0}\sin\varphi \sum_{i=1}^{i=f_s \cdot T} \dot{X}_i^2 t_i \tag{4}$$

where f_s refers to the sampling frequency of the experiment, f_s = 1024 Hz.

Due to the equal-interval sampling of the experiment data, $t_i = \frac{1}{1024}$.

2.2. Series Model

As shown in Figure 3, the standard equation of the model is:

$$\begin{cases} M\ddot{X} + k(X - x) = F \\ k(x - X) + c\dot{x} = 0 \end{cases} \tag{5}$$

Substitute F=F0eiωt and X=X0ei(ωt−φ) into the equation and apply Euler's Formula, eiφ=cosφ+isinφ, to determine the damping and stiffness:

$$c = \frac{F_0}{\omega X_0}\sin\varphi + \frac{F_0}{\omega X_0}\cot\varphi\cos\varphi + 2\omega M\cot\varphi + \frac{\omega^3 M^2 X_0}{F_0\sin\varphi} \tag{6}$$

$$k = \frac{c}{\frac{1}{\omega}\cot\varphi + \frac{\omega M X_0}{F_0\sin\varphi}} \tag{7}$$

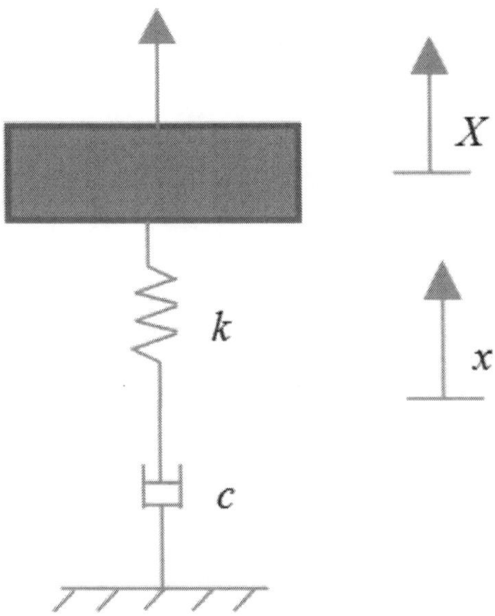

Figure 3. Series model.

The periodical energy consumption of damping is:

$$q = \int c\dot{x}\mathrm{d}x = \int_0^T c\dot{x}^2\mathrm{d}t = \sum_{i=1}^{i=f_s\cdot T} c\dot{x}_i^2 t_i \tag{8}$$

2.3. Work Done by Excitation Force

The periodical work done by excitation force of the system is:

$$W = \int F \mathrm{d}X = \int_{0}^{T} F\dot{X}\mathrm{d}t = \sum_{i=1}^{i=f_s \cdot T} F_i \dot{X}_i t_i \quad (9)$$

where F refers to the excitation load, and X refers to the structural response.

3. EXPERIMENTAL METHOD

3.1. Design of Experimental Device

The dual-chamber air spring vibration absorber described here is mainly used to restrain the large vibration of an offshore structure. In the experiment, a large water container was designed to simulate the action of sea water. The container is a cylindrical device with a diameter of 1500 mm (much larger than the diameter of 100 mm of the air chamber). As shown in Figure 4, the experimental device consists of: 1. Hydraulic pressure loader; 2. Force sensor; 3. Dual-chamber air spring; 4. Orifice plate; and 5. Air pressure sensor.

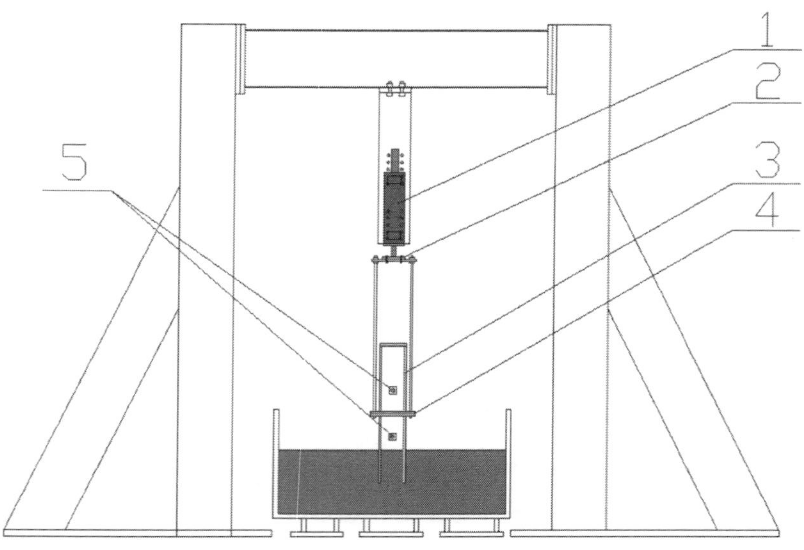

Figure 4. Schematic diagram for the experimental device.

As shown in Figure 5, in order to study the effects of different upper and lower air chamber ratios on the vibrating characteristics, the length of the lower air chamber (200 mm) was kept constant while the upper air chamber was changed （100 mm, 150 mm, 200 mm, 250 mm, 300 mm, 350 mm, 400 mm） to study the effect of different volume ratios. The upper and lower air chambers are connected by an orifice plate and the air flows back and forth through an orifice. The orifice changes the air distribution of the chambers greatly and can influence the vibrating characteristics of the structure. In this experiment, the orifice diameters used to study the effect of different orifice ratios on the air spring were 0, 10, 20, 30, 40, 50, and 60 mm.

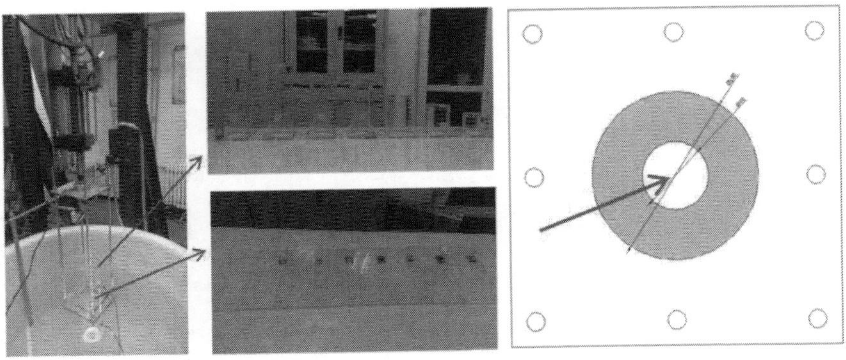

Figure 5. Upper chamber and orifice plates of the dual-chamber air spring.

As shown in Figure 6, a hydraulic servo fatigue machine is used as the loading system of the experiment. The fatigue machine has excellent performance and can be used for large structure experiments. By controlling the loading frequency and amplitude, we can test the effect of external load on the vibrating characteristics of the air spring. In the experiment, the amplitude range was 2, 4, 6, 8, 10, 12, and 14 mm, and the frequency range was 0.497–5.474 Hz.

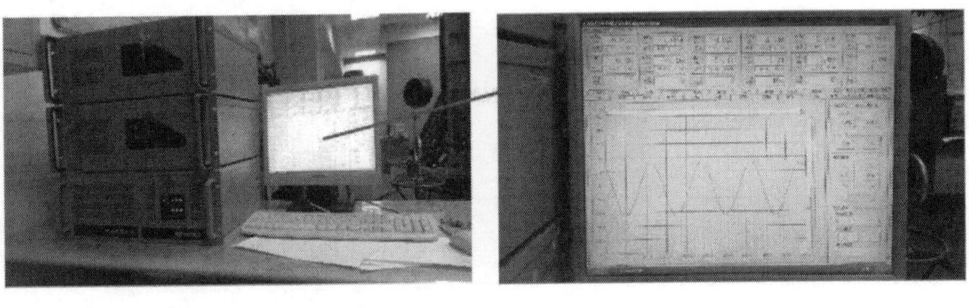

Figure 6. Loading system for experimental device.

3.2. Setting of Test Parameters

We studied the influences of each factor on dual-chamber air spring dynamic stiffness and damping. The experiment assumes the cross section of the upper air chamber equals that of the lower air chamber.

Variables to be considered: Volume of upper air chamber V_2, Volume of lower air chamber V_1, opening diameter d, amplitude of outer excitation A, and frequency of outer excitation ω.

The length of the lower air chamber, L_1, and the frequency of outer excitation, ω, were used as the basic variables with the following three dimensionless numbers:

Height of lower air chamber $L_1 = 0.2$ m

Diameter of air chamber $D = 0.1$ m

$\xi_1 = \frac{L_2}{L_1}$, volume ratio, values = 0.5, 0.75, 1, 1.25, 1.5, 1.75, 2

$\xi_2 = \frac{d}{D}$, orifice ratio, values = 0, 0.1, 0.2, 0.3, 0.4, 0.5, 0.6

$\xi_3 = \frac{A}{L_1}$, ratio of the amplitude of outer excitation and the height of air chamber, values = 0.01, 0.02, 0.03, 0.04, 0.05, 0.06, 0.07.

Frequency of outer excitation. Log value $\ln\omega$ was used to investigate the vibrating characteristics of the air spring at low frequency (values = −0.7, −0.3, 0.1, 0.5, 0.9, 1.3, 1.7).

In total, four variables and seven values were tested, as shown in Table 1:

Table 1. Four factors and seven levels chart.

Level / Factor	Volume Ratio (ξ_1)	Orifice Ratio (ξ_2)	Loading Amplitude Ratio (ξ_3)	Loading Frequency $\ln\omega$	Real Loading Frequency (Hz)
1	0.5	0.0	0.01	−0.7	0.497
2	0.75	0.1	0.02	−0.3	0.741
3	1.0	0.2	0.03	0.1	1.105
4	1.25	0.3	0.04	0.5	1.649
5	1.5	0.4	0.05	0.9	2.460
6	1.75	0.5	0.06	1.3	3.669
7	2.0	0.6	0.07	1.7	5.474

3.3. Orthogonal Experiment and Orthogonal Array

For experimental designs with multiple factors and multiple levels, a large number of experiments are required. For example, for the experimental scheme with four variables and seven levels in this paper, 2401 (7^4) experiments need to be carried out. Hence, a method with the minimum numbers of experiments is needed to find out the effect of each parameter on the vibrating characteristics of the air spring.

In 1947, Rao [31] developed factorial experiments, an optimization experimental method (an efficient testing strategy) that can determine the relation between factors and a test index and specify the primary and secondary factors using fewer trials. Taguchi [32] effectively used orthogonal arrays in his research. Azouzi and Guillot [33] examined the feasibility for an intelligent sensor fusion to estimate online surface finish and dimensional deviations using orthogonal arrays. Green, Krieger, and Wind [34] applied orthogonal arrays to conjoint analysis. The use of the orthogonal test method minimizes the number of tests but still allows determination of the changing rules of all factors. We applied the orthogonal experiment method to investigate the effect of upper and lower air chamber volume ratio, orifice ratio, loading amplitude, and loading frequency of outer excitation.

As an effective method of solving multi-factor experiment problems, orthogonal experimental design selects typical points from the overall experiment and tests them. These typical points are "even" and "regular": (1) in the experiment, each factor has the same occurrence number at each different level; (2) each combination of two factors at each different level occurred in the experiment and had the same occurrence number. Meeting these two criteria indicates that the experiment scheme designed by orthogonal experiment method is typical. This method reduces the number of experiments and can generally reflect the overall effect of each factor at each level on the index.

Range analysis is commonly used in scientific experiments. Change in experimental results often occurs in scientific experiments due to two types of factors. One is variations arising from random effects; such effects are controllable in experiments, and are therefore inevitable. The other is artificial control produces changes in experimental results. When such factors have a significant impact on the experiment, they are bound to significantly alter the results, accompanied by random factors. Conversely, when such factors have no significant effect on the experimental results, corresponding changes will not be manifested obviously, so changes in experimental results have been substantially ascribed to the effects of random factors. The purpose of conducting scientific experiments is often to determine whether these artificially controllable factors have an impact on the experimental results and what the effect is.

Range analysis is an effective tool to judge the above matters through analyzing the data variation in experimental results, because it can separate random variation from non-random variation in a hybrid state to help determine the source of dominant variations. Using range analysis, we analyzed the results of orthogonal design to explore the effect of various factors on air spring characteristics. In addition, we investigated the effects of four factors (upper and lower air chamber volume ratio, orifice ratio, loading amplitude, and loading frequency of outer excitation) on the air vibration characteristics of the dual-chamber. Seven variation levels of each factor were examined. The tests were performed 49 times, accounting for only 2% of the traditional number of trials. The arrangements and results of the orthogonal analysis scheme are shown in Appendix 1.

4. EXPERIMENTAL RESULTS AND ANALYSIS OF A DUAL-CHAMBER AIR SPRING

We used an intuitive method in orthogonal analysis that looks at means and ranges. The mean average is the average value of different factors at the same level, and this can reveal the effect of different levels of single factors on the indicators. The range indicates the maximum value of the average numerical difference for each factor at each level, and reflects the impact of each column factor at different levels, from which the primary and secondary sequence of factors can be determined. If each factor has the same number of level, the influence extent of each factor can be judged directly by comparing the range size.

For 49 sets of experimental schemes, the comparison of energy dissipation is shown in Figure 7.

Figure 7 shows that under most operating conditions, there is a good fit between the parallel model and work done by excitation force, but under individual conditions, the series model shows better goodness of fit. From the relationship of energy dissipation mean of the 49 sets of orthogonal design schemes, the goodness of fit of the parallel model $\alpha_p = 1 - \frac{\sum_{0}^{49} \frac{\left|Q_{parallel} - Q_{structure}\right|}{Q_{structure}}}{49} = 99.88\%$.

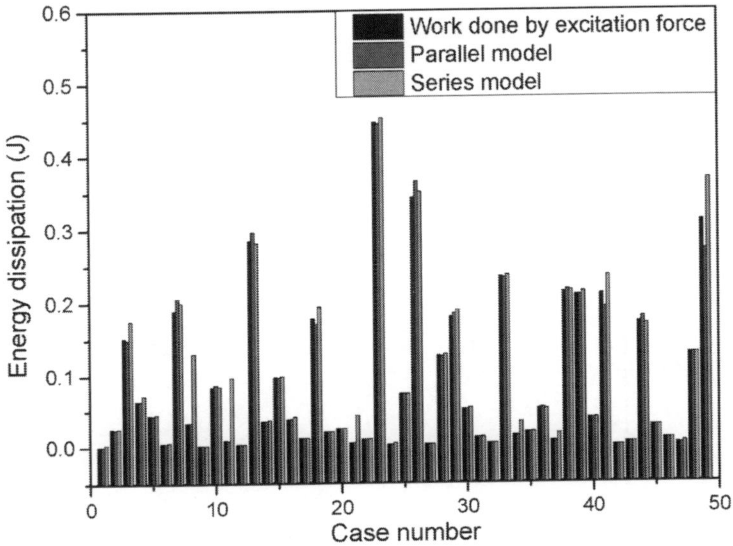

Figure 7. Energy dissipation test program comparison chart.

The goodness of fit of the series model $\alpha_s = 1 - \frac{\sum_{0}^{49} \frac{|Q_{series} - Q_{structure}|}{Q_{structure}}}{49} = 89.43\%.$.

By the range relation of energy consumption, we can infer a higher goodness of fit between work done by excitation force and energy dissipation of the parallel model. Furthermore, the parallel model can be employed to make deeper analyses of the structure.

4.1. Influence of Various Parameters on Air Spring Stiffness

Stiffness is a function of the volume ratio, orifice ratio, excitation amplitude, and frequency. The influence law can be seen from the following figures:

As can be seen from Figure 8, the effect of the volume ratio on the dual-chamber vibration absorber stiffness is not monotonic. With an increase in the volume ratio, the stiffness of vibration absorber first slowly decreases and increases at a volume ratio of 1.25, then decreases and finally increases dramatically. As per Figure 9, the effect of the orifice ratio on the dual-chamber vibration absorber stiffness similarly is not monotonic. With an increase in the orifice ratio, the stiffness of vibration absorber first shows little change, then increases at a orifice ratio of 0.4, then decreases and finally increases to its highest level.

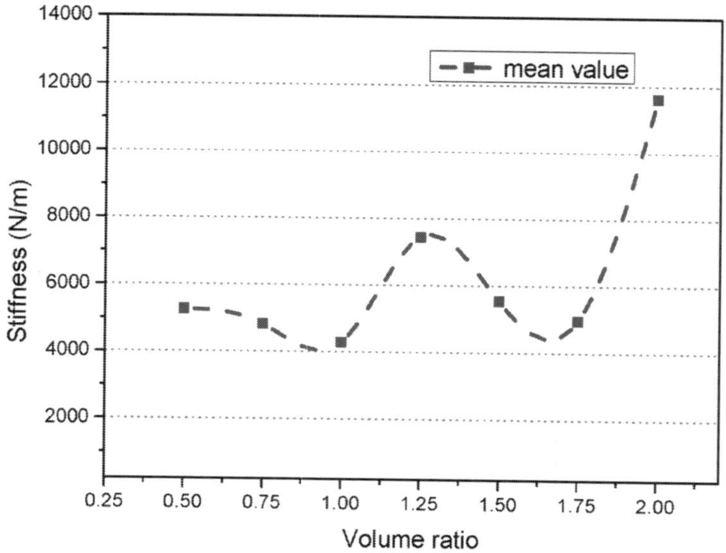

Figure 8. Effect of the volume ratio on the dual-chamber air spring stiffness.

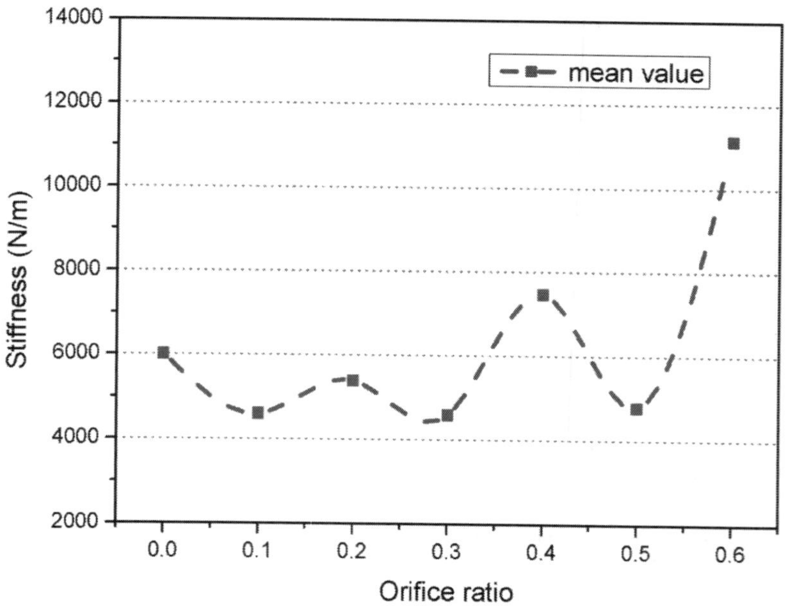

Figure 9. Effect of the orifice ratio on the dual-chamber air spring stiffness.

According to Figure 10, the effect of the loading amplitude ratio on the dual-chamber vibration absorber stiffness is basically monotonic. With an increase in the amplitude ratio, the stiffness of the dual-chamber air spring also increases and slightly protrudes at the amplitude ratio of 0.04. Figure 11 shows the effect of the loading frequency on the dual-chamber vibration absorber stiffness is also monotonic. The stiffness of the dual-chamber air spring monotonically increases with an increase in the loading frequency. As can be seen from Figure 12, the effect of each factor on the dual-chamber vibration absorber stiffness varies, where the volume ratio, orifice ratio, and loading amplitude ratio exert considerable influence but the loading frequency has the maximum impact. The results described in Figure 12 are only applicable when considering the volume ratio, the orifice ratio, load amplitude, and frequency in this case.

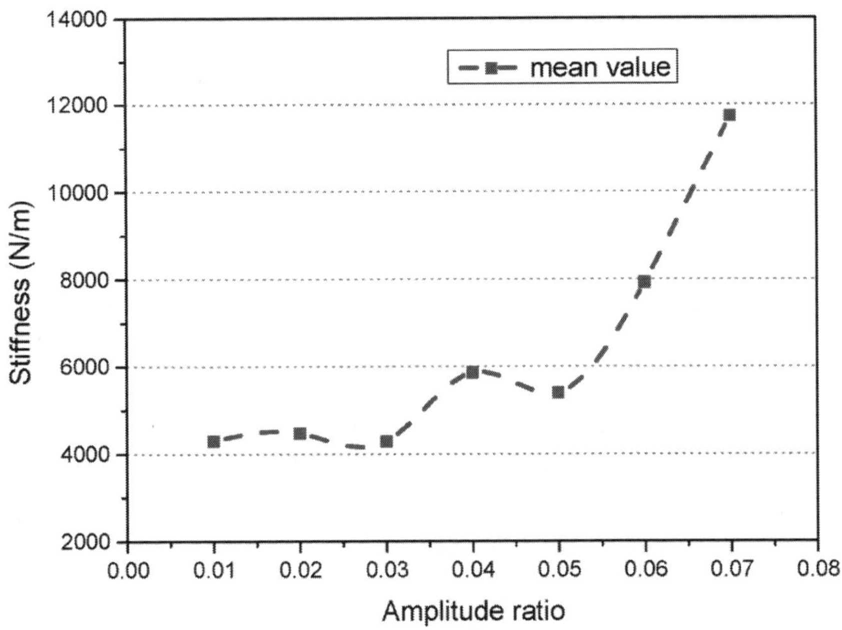

Figure 10. Effect of the amplitude ratio on the dual-chamber air spring stiffness.

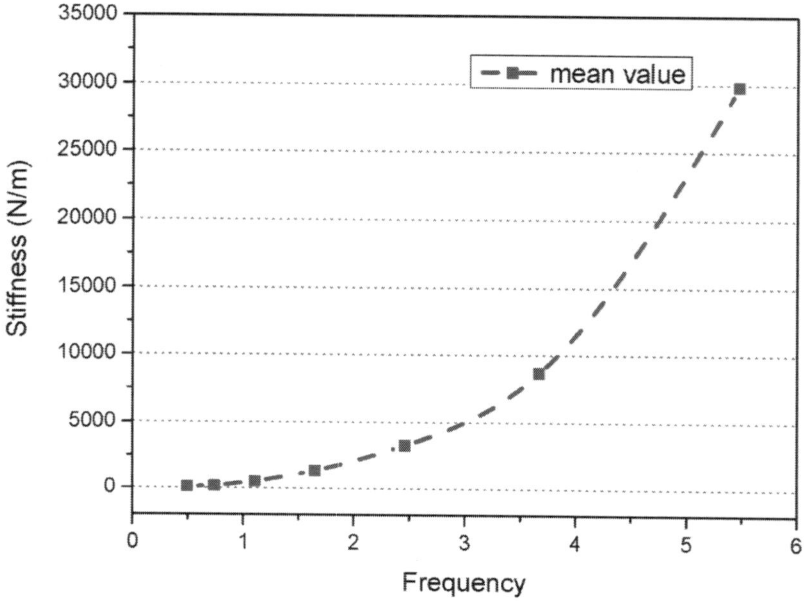

Figure 11. Effect of the frequency on the dual-chamber air spring stiffness.

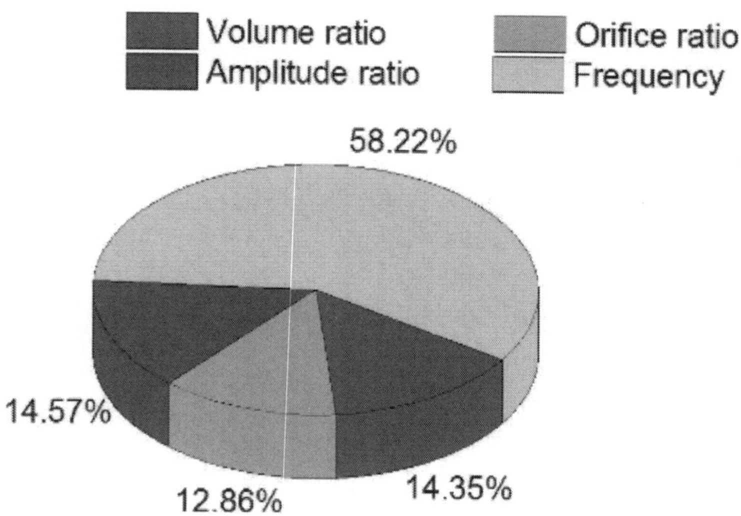

Figure 12. Summary of the effects of the four factors on the dual-chamber air spring stiffness.

4.2. Influence of Various Parameters on Air Spring Damping

Damping is a function of the volume ratio, orifice ratio, excitation amplitude, and frequency. In this section the four factors for the dual-chamber air spring damping effects are examined.

The influence law can be seen from the following figures:

Figure 13 shows that the effect of the volume ratio of the dual-chamber vibration absorber damping is not monotonic. With an increase in the volume ratio, the damping of the vibration absorber first slowly decreases and then increases at a volume ratio of 1.25, then decreases and then increases. As per Figure 14, the effect of the orifice ratio on the dual-chamber vibration absorber damping is also not monotonic. With an increase in the orifice ratio, the damping of vibration absorber first shows little change, then increases at an orifice ratio of 0.4, then decreases and then increases.

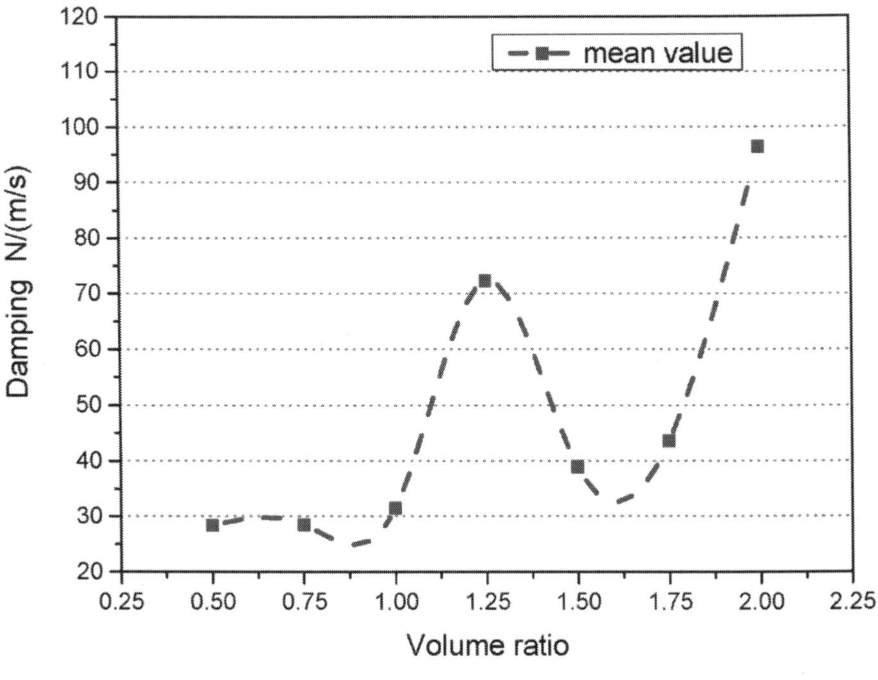

Figure 13. Effect of the volume ratio on the dual-chamber air spring damping.

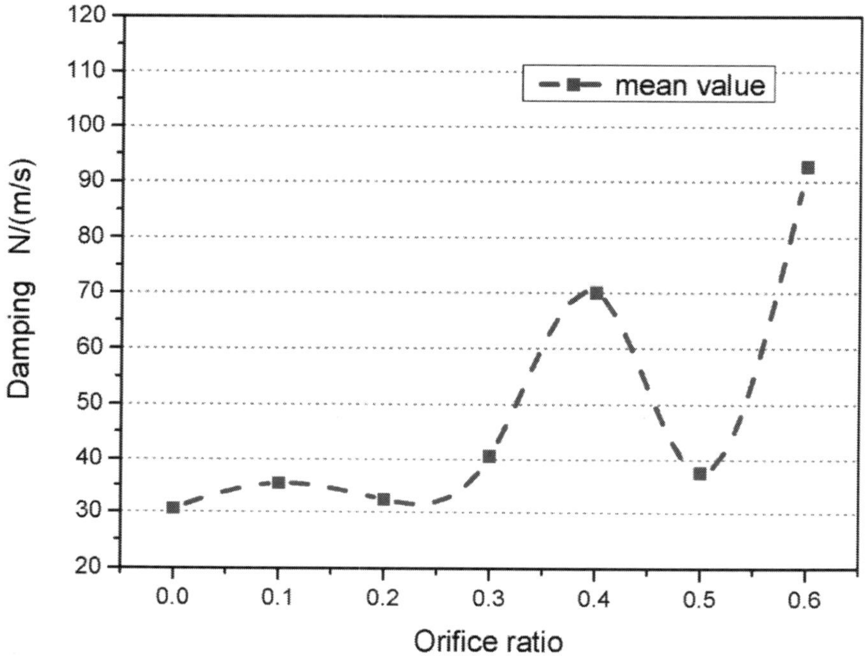

Figure 14. Effect of the orifice ratio on the dual-chamber air spring damping.

According to Figure 15, the effect of the loading amplitude ratio on the dual-chamber vibration absorber damping is not monotonic. With an increase in the amplitude ratio, the structural damping shows little change at first and then declines at 0.05, before eventually increasing. Figure 16 indicates that the effect of the loading frequency on the dual-chamber vibration absorber damping is basically monotonic. With an increase in the amplitude ratio, the vibration absorber damping increases until 3.669 Hz, and then increases dramatically. As summarized in Figure 17, the effect of each factor on the dual-chamber vibration absorber damping varies, where the volume ratio, orifice ratio, and loading amplitude ratio have strong effects but the loading frequency has the maximum impact. The results described in Figure 17 are only applicable when considering the volume ratio, the orifice ratio, load amplitude, and frequency in this case.

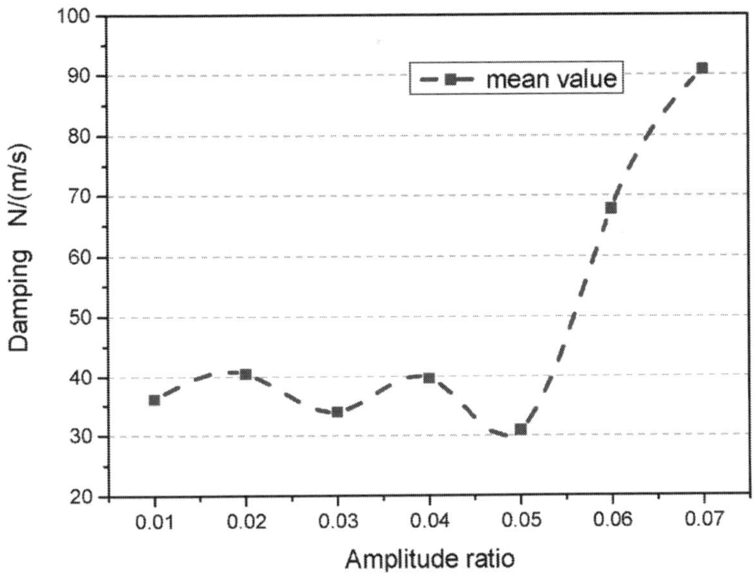

Figure 15. Effect of the amplitude ratio on the dual-chamber air spring damping.

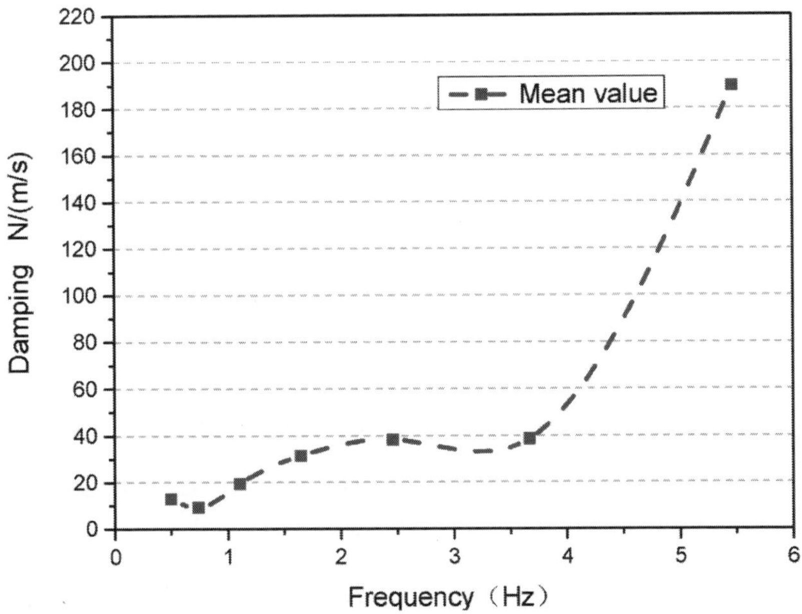

Figure 16. Effect of the frequency on the dual-chamber air spring damping.

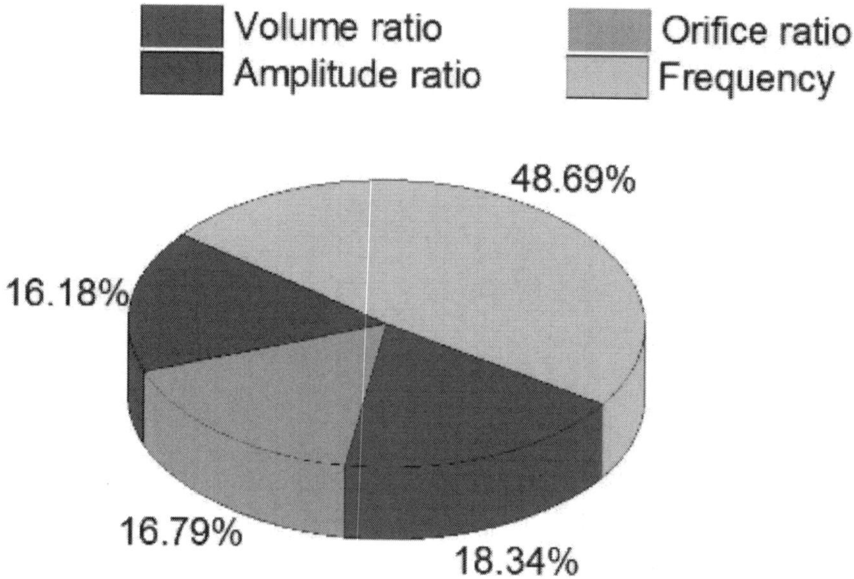

Figure 17. Summary of the effects of the four factors on dual-chamber air spring damping.

Thus, the four factors influence the stiffness and damping of the gas-liquid coupling dual-chamber air spring in a complex manner; there is not a simple linear relationship. Each factor has different effects on stiffness and damping. Prediction of the effects on stiffness and damping of the gas-liquid coupling dual-chamber air spring based on these four factors would be useful for efficient design, and in the next chapter, we propose an efficient model of dual-chamber air spring stiffness and damping to do this.

5. RAPID CALCULATION MODEL OF DUAL-CHAMBER AIR SPRING STIFFNESS AND DAMPING

From the previous analyses, we can see that volume ratio, orifice ratio, loading amplitude, and loading frequency all have complex effects on the stiffness and damping of the dual-chamber air spring device and are not monotonic in most cases. Therefore complex models must be constructed using statistical analyses of the experimental data.

5.1. Normalization Process of Each Factor

In order to facilitate the analysis of each factor's effect, the normalization process of each factor is performed.

Normalized volume ratio $\varphi_1 = \frac{\xi_1}{2}$, values of 0.25, 0.375, 0.5, 0.625, 0.75, 0.875, 1.

Normalized orifice ratio $\varphi_2 = \frac{\xi_2}{0.6}$, values of 0, 0.167, 0.333, 0.5, 0.667, 0.833, 1.

Normalized loading amplitude ratio $\varphi_3 = \frac{\xi_3}{0.07}$, values of 0.142, 0.286, 0.429, 0.571, 0.714, 0.857, 1.

Normalized loading frequency $\varphi 4 = \ln\omega + 0.72.4$, values of 0, 0.167, 0.333, 0.5, 0.667, 0.833, 1.

The functional relationship between the stiffness and damping of the dual-chamber air spring and each factor is:

$$K_{\text{modle}} = f_1(\varphi_1, \varphi_2, \varphi_3, \varphi_4) \quad (10)$$

$$C_{\text{modle}} = f_2(\varphi_1, \varphi_2, \varphi_3, \varphi_4) \quad (11)$$

5.2. Determination of Rapid Calculation Model of Stiffness and Damping

From the above analysis, each factor has a nonlinear impact on the stiffness and damping of the dual-chamber air spring. The quadratic function can not accurately simulate the variation law of the stiffness and damping of the dual-chamber air spring. When the power of an independent variable is more than three, the regression function becomes very unstable [35], thus destabilizing the application of the regression model. In order to more accurately reflect the effect law of each factor on the stiffness and damping of the dual-chamber air spring, this paper uses cubic polynomials, which can usually be converted into ordinary multiple linear regression for processing. Experimental data are used to fit the expression of the stiffness and damping of the dual-chamber air spring.

Suppose the stiffness experimental value meets the following cubic polynomials:

$$\begin{aligned}
f_1 &= \beta_0 + \sum_{j=1}^{34} \beta_j g_j(\varphi_1, \varphi_2, \varphi_3, \varphi_4) \\
&= \beta_0 + \beta_1\varphi_1 + \beta_2\varphi_2 + \beta_3\varphi_3 + \beta_4\varphi_4 + \beta_5\varphi_1^2 + \beta_6\varphi_2^2 + \beta_7\varphi_3^2 + \beta_8\varphi_4^2 + \beta_9\varphi_1\varphi_2 + \beta_{10}\varphi_1\varphi_3 + \beta_{11}\varphi_1\varphi_4 \\
&+\beta_{12}\varphi_2\varphi_3 + \beta_{13}\varphi_2\varphi_4 + \beta_{14}\varphi_3\varphi_4 + \beta_{15}\varphi_1^3 + \beta_{16}\varphi_2^3 + \beta_{17}\varphi_3^3 + \beta_{18}\varphi_4^3 + \beta_{19}\varphi_1^2\varphi_2 + \beta_{20}\varphi_1^2\varphi_3 + \beta_{21}\varphi_1^2\varphi_4 \\
&+\beta_{22}\varphi_1\varphi_2^2 + \beta_{23}\varphi_1\varphi_3^2 + \beta_{24}\varphi_1\varphi_4^2 + \beta_{25}\varphi_2^2\varphi_3 + \beta_{26}\varphi_2^2\varphi_4 + \beta_{27}\varphi_2\varphi_3^2 + \beta_{28}\varphi_2\varphi_4^2 + \beta_{29}\varphi_3^2\varphi_4 + \beta_{30}\varphi_3\varphi_4^2 \\
&+\beta_{31}\varphi_1\varphi_2\varphi_3 + \beta_{32}\varphi_1\varphi_2\varphi_4 + \beta_{33}\varphi_1\varphi_3\varphi_4 + \beta_{34}\varphi_2\varphi_3\varphi_4 + \varepsilon_i(o^4)
\end{aligned} \tag{12}$$

where β0,β1,β2,....,β34 are undetermined coefficients; φ1,φ2,φ3,φ4 are controllable variables in the experiment; gj(φ1,φ2,φ3,φ4)is function expression corresponding to βj; and f1 is the stiffness of dual-chamber air spring.

This paper uses the least squares method to estimate the undetermined coefficients in the above formula to obtain the regression equation for stiffness:

$$\begin{aligned}
\frac{\hat{f}_1}{E13} &= -0.0952 + 1.4139\varphi_1 - 0.6852\varphi_2 - 1.3314\varphi_3 + 1.2366\varphi_4 - 0.9296\varphi_1^2 + 0.5123\varphi_2^2 \\
&+0.3488\varphi_3^2 - 0.7531\varphi_4^2 - 0.3542\varphi_1\varphi_2 - 0.7975\varphi_1\varphi_3 - 1.5502\varphi_1\varphi_4 + 2.7841\varphi_2\varphi_3 \\
&-0.5386\varphi_2\varphi_4 + 0.2573\varphi_3\varphi_4 - 0.1684\varphi_1^3 - 0.5071\varphi_2^3 + 1.4815\varphi_3^3 - 0.2310\varphi_4^3 \\
&+0.8761\varphi_1^2\varphi_2 + 1.3092\varphi_1^2\varphi_3 + 0.1184\varphi_1^2\varphi_4 - 0.0996\varphi_1\varphi_2^2 - 0.3868\varphi_1\varphi_3^2 + 0.0888\varphi_1\varphi_4^2 \\
&+1.1500\varphi_2^2\varphi_3 - 0.6931\varphi_2^2\varphi_4 - 2.5784\varphi_2\varphi_3^2 + 0.1512\varphi_2\varphi_4^2 - 2.7151\varphi_3^2\varphi_4 + 1.6950\varphi_3\varphi_4^2 \\
&-2.1074\varphi_1\varphi_2\varphi_3 + 1.1258\varphi_1\varphi_2\varphi_4 + 1.3134\varphi_1\varphi_3\varphi_4 + 0.6595\varphi_2\varphi_3\varphi_4
\end{aligned} \tag{13}$$

Similarly, the damping regression equation can be acquired based on the stiffness calculation method:

$$\begin{aligned}
\frac{\hat{f}_2}{E11} &= -0.2648 + 3.0819\varphi_1 - 1.6038\varphi_2 - 2.5479\varphi_3 + 2.5604\varphi_4 - 2.1802\varphi_1^2 + 1.4020\varphi_2^2 \\
&+1.5283\varphi_3^2 - 1.6530\varphi_4^2 - 0.8512\varphi_1\varphi_2 - 1.5508\varphi_1\varphi_3 - 3.7226\varphi_1\varphi_4 + 5.4696\varphi_2\varphi_3 \\
&-0.8495\varphi_2\varphi_4 + 0.0787\varphi_3\varphi_4 - 0.1734\varphi_1^3 - 0.9562\varphi_2^3 + 1.4337\varphi_3^3 - 0.2797\varphi_4^3 \\
&+1.6602\varphi_1^2\varphi_2 + 2.1616\varphi_1^2\varphi_3 + 0.8800\varphi_1^2\varphi_4 - 0.1481\varphi_1\varphi_2^2 - 0.3982\varphi_1\varphi_3^2 + 0.6600\varphi_1\varphi_4^2 \\
&+0.9527\varphi_2^2\varphi_3 - 0.8391\varphi_2^2\varphi_4 - 3.9493\varphi_2\varphi_3^2 + 0.4225\varphi_2\varphi_4^2 - 3.5250\varphi_3^2\varphi_4 + 2.5354\varphi_3\varphi_4^2 \\
&-3.3547\varphi_1\varphi_2\varphi_3 + 1.6821\varphi_1\varphi_2\varphi_4 + 1.9625\varphi_1\varphi_3\varphi_4 + 0.3758\varphi_2\varphi_3\varphi_4
\end{aligned} \tag{14}$$

Then the stiffness and damping of the dual-chamber air spring can be calculated by the following formulas:

$$K_{modle} = f_1(\varphi_1, \varphi_2, \varphi_3, \varphi_4) = \hat{f}_1 \cdot E13 \tag{15}$$

$$C_{\text{modle}} = f_2(\varphi_1, \varphi_2, \varphi_3, \varphi_4) = \hat{f}_2 \cdot E11 \tag{16}$$

Equations (15) and (16) determine the cubic polynomial equation of stiffness and damping.

5.3. Test of Calculation Model

In statistics, regression analysis is performed using the variables. The coefficient of determination. R^2 is the ratio between the regression sum of squares and sum of squares for total. R^2 is between 0 and 1. The closer it is to 1, the better the fit of the regression. A goodness of fit of more than 0.8 is typically considered acceptable.

In multiple regression analysis, R^2 expression:

$$R^2 = 1 - \frac{SSE}{SST} \tag{17}$$

where SST is the sum of squares for total; and SSE is the sum of squares for error.

$$SST = \sum_{i=1}^{N} \left(f_{1i} - \bar{f}_{1i}\right)^2 \tag{18}$$

$$SSE = \sum_{i=1}^{N} \left(f_{1i} - \hat{f}_{1i}\right)^2 \tag{19}$$

The coefficient of determination R^2 is related to the number of independent variables. In order to accurately test the accuracy of the model, it is necessary to take into account the degrees of freedom and make adjustments of R^2 expression according to the size of the formula.

$$R_a^2 = 1 - \frac{SSE/(n-p-1)}{SST/(n-p)} = 1 - \left(1 - R^2\right) \cdot \left(\frac{n-1}{n-p-1}\right) \tag{20}$$

R_a^2 is denoted as the adjusted coefficient of determination; n is the total number of experiments, and p is the number of variables. The adjusted coefficient of determination more accurately reflects the degree of fit between the model and experimental data.

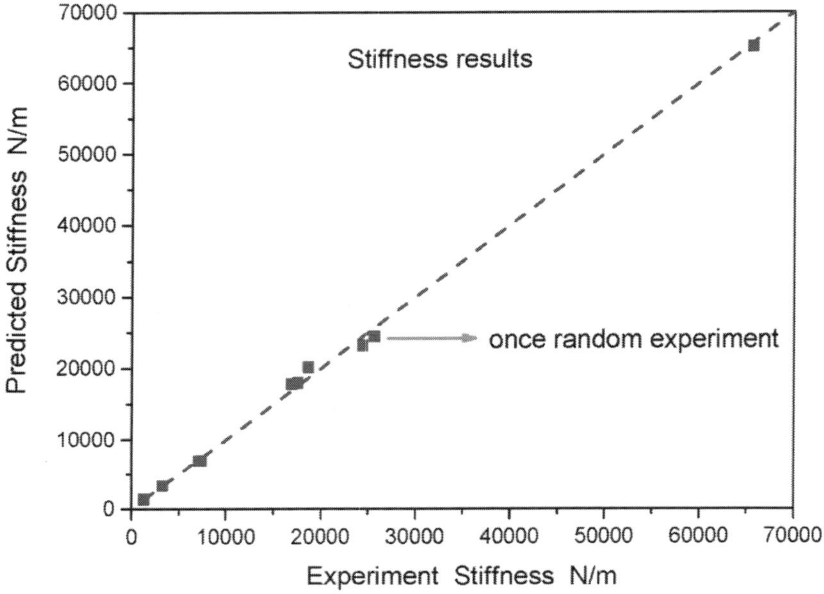

Figure 18. Comparison of experimental and predicted values on the dual-chamber air spring stiffness.

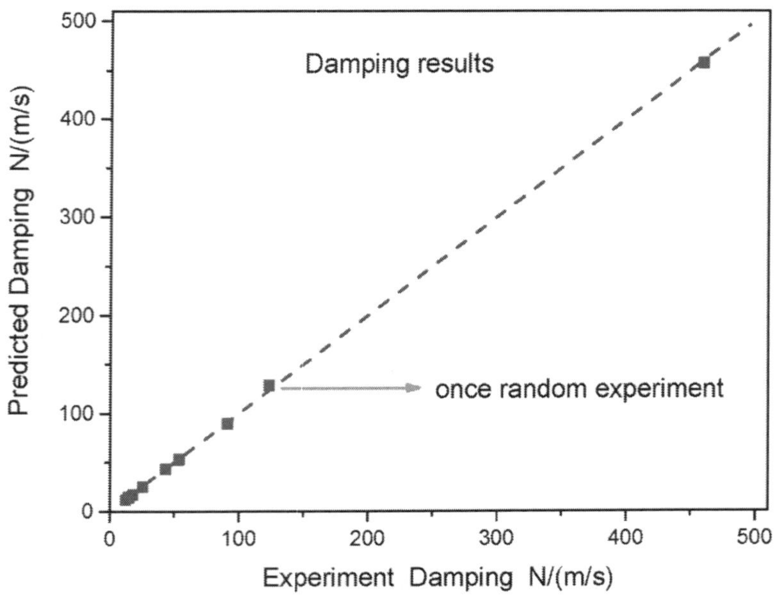

Figure 19. Comparison of experimental and predicted values on the dual-chamber air spring damping.

By making an analysis of the data ($n = 49, p = 35$), the following can be obtained:

1. For the stiffness model $R^2 = 0.99154$, $R_a^2 = 0.96378$, the goodness of fit is 96.378%, indicating that the model can well simulate and predict the experimental results. This indicates the effectiveness of the cubic polynomial regression equation fit.
2. For the damping model, $R^2 = 0.98970$, $R_a^2 = 0.95470$, the goodness of fit is 95.470%, again indicating that the model can well simulate and predict the experimental results. This again supports the effectiveness of the cubic polynomial regression equation fit.

5.4. Prediction of Experimental Results by the Rapid Calculation Model

To further validate the forecast accuracy of the stiffness and damping rapid calculation model, we conducted a series of experiments. To ensure the universality, these experiments consist of random combinations of four factors at different levels, different from the levels tested in the test programs in the orthogonal table.

The straight lines in Figure 18 and Figure 19 are experimental values, while red dots are predictive values for the rapid calculation model. If a dot falls on the line, this indicates the predicted value is equivalent to the experimental value; deviations from the straight line indicate differences between the predicted and the experimental values. As seen in the figures, the predicted and experimental values fit well. For more horizontal level combination schemes, the predicted value can be calculated based on the rapid calculation model proposed in this paper, which can play a guiding role to engineering structural design and applications.

The experiments performed here were representative in the parameter range, and the polynomial prediction model is appropriate for these values. The application of the polynomial prediction model is as follows: upper and lower air chamber volume ratio, $\xi 1 \in [0.5–2.0]$, orifice ratio, $\xi 2 \in [0–0.6]$, loading amplitude ratio, $\xi 3 \in [0.01–0.07]$, and loading frequency $f \in [0.497–5.474]$.

6. CONCLUSIONS

The mechanical differences between the gas-liquid coupling dual-chamber air spring device and the traditional one make the analysis of effects on stiffness and damping more challenging. The compression of the gas in the traditional air spring is directly on the external load of the air chamber by the piston. The pressure variation of the air spring chamber is the same phase and amplitude as the external load, but for the dual-chamber air spring device applied to motion suppression of marine structures, the movement of the liquid column and the external excitation differ in phases and amplitude.

The current study analyzes the characteristics of a dual-chamber air spring device. A parallel model and series model were used to simulate the dual-chamber air spring energy-absorbing device. An orthogonal test scheme was used to investigate the effect laws of four factors (upper and lower air chamber volume ratio, orifice ratio, loading amplitude, and loading frequency of outer excitation) on the air vibration characteristics of the dual-chamber. Based on the experimental results, a higher-order non-linear regression method was obtained, achieving a rapid calculation model for dual-chamber air spring stiffness and damping, and the reliability of the method was verified experimentally. Using the rapid calculation model, from the upper and lower air chamber volume, orifice ratio, the frequency, and amplitude of external load, we can determine the stiffness and damping of the dual-chamber air spring device applied to motion suppression of marine structures by the formulas (15) and (16). The main findings are:

1. Based on energy consumption results, the goodness of fit of the parallel model was 89.43%, and the goodness of fit of the series model was 99.88%. The parallel model is more consistent with the real physical model.

2. The effects of volume ratio and orifice ratio on dual-chamber vibration absorber stiffness were not monotonic, but the loading amplitude ratio and frequency tended toward monotonic increasing.

3. The effects of the volume ratio, orifice ratio, and loading amplitude ratio on the dual-chamber vibration absorber stiffness did not behave in a monotonic manner, but the loading frequency on damping tended toward monotonic increasing.

4. A polynomial rapid calculation model for stiffness and damping was constructed. The accuracy of the rapid calculation model

results was verified by the experimental results, and the predicted values were in good agreement with the experimental values.

ACKNOWLEDGMENTS

The authors would like to express their gratitude for the financial support of the National Natural Science Foundation of China (Grant Nos. 11072246, 51490673, 10702073, 11232012) and the National Basic Research Program of China (973 Program 2014CB046801)

AUTHOR CONTRIBUTIONS

Xiaohui Zeng put forward the overall idea. Liang Zhang contributed to the theoretical and experimental analysis, and wrote the manuscript. Yang Yu contributed to the experimental study. Min Shi contributed to data analysis. Jifu Zhou contributed to data analysis.

CONFLICTS OF INTEREST

The authors declare no conflict of interest.

REFERENCES

1. Taylor, R.E.; Jefferys, E.R. Variability of hydrodynamic load predictions for a tension leg platform. *Ocean Eng.* **1986**, *13*, 449–490.
2. Zeng, X.H.; Shen, X.P.; Wu, Y.X. Governing equations and numerical solutions of tension leg platform with finite amplitude motion. *Appl. Math. Mech.* **2007**, *28*, 37–49.
3. Zeng, X.H.; Li, X.W.; Liu, Y.; Wu, Y.X. Nonlinear dynamic responses of tension leg platform with slack-taut tether. *China Ocean Eng.* **2009**, *23*, 37–48.
4. Ahmad, S.K.; Ahmad, S. Active control of non-linearly coupled TLP response under wind and wave environments. *Comput. Struct.* **1999**, *72*, 735–747.

5. Alves, R.M.; Batista, R.C. *Active/Passive Control of Heave Motion for TLP Type Offshore Platform*; Chung, J.S., Matsui, T., Koterayama, W., Eds.; International Society of Offshore and Polar Engineers: Cupertino, CA, USA, 1999; pp. 332–338.

6. Sakai, F.; Takaeda, S.; Tamaki, T. Tuned liquid column damper-new type device for suppression of building vibrations. In Proceedings of the International Conference on Highrise Buildings, Nanjing, China, 25–27 March 1989; pp. 926–931.

7. Lee, H.H.; Wong, S.H.; Lee, R.S. Response mitigation on the offshore floating platform system with tuned liquid column damper. *Ocean Eng.* **2006**, *33*, 1118–1142.

8. Lee, H.H.; Juang, H.H. Experimental study on the vibration mitigation of offshore tension leg platform system with UWTLCD.*Smart Struct. Syst.* **2012**, *9*, 71–104.

9. Taflanidis, A.A.; Angelides, D.C.; Scruggs, J.T. Simulation-based robust design of mass dampers for response mitigation of tension leg platforms. *Eng. Struct.* **2009**, *31*, 847–857.

10. Lee, S.K.; Lee, H.R.; Min, K.W. Experimental verification on nonlinear dynamic characteristic of a tuned liquid column damper subjected to various excitation amplitudes. *Struct. Des. Tall Spec. Build.* **2012**, *21*, 374–388.

11. Chatterjee, T.; Chakraborty, S. Vibration mitigation of structures subjected to random wave forces by liquid column dampers.*Ocean Eng.* **2014**, *87*, 151–161.

12. Zeng, X.; Yu, Y.; Zhang, L.; Liu, Q.; Wu, H. A New Energy-Absorbing Device for Motion Suppression in Deep-Sea Floating Platforms. *Energies* **2014**, *8*, 111–132.

13. Rijken, O.; Spillane, M.; Leverette, S.J. Vibration Absorber Technology and Conceptual Design of Vibration Absorber for TLP in Ultradeep Water. In Proceedings of the ASME 29th International Conference on Ocean, Offshore and Arctic Engineering, Shanghai, China, 6–11 June 2010; ASME: New York, NY, USA, 2010; pp. 629–638.

14. Bian, X.S.; Leverette, S.J.; Rijken, O.R. A TLP solution for 8000 Ft water depth. In Proceedings of the ASME 29th International Conference on Ocean, Offshore and Arctic Engineering, Shanghai, China, 6–11 June 2010; ASME: New York, NY, USA, 2010; pp. 255–262.

15. Spillane, M.W.; Rijken, O.R.; Leverette, S.J. Vibration absorbers for deepwater TLP's. In Proceedings of the 17th International Offshore and Polar Engineering Conference, Lisbon, Portugal, 1–6 July 2007.

16. Bachrach, B.I.; Rivin, E. Analysis of a damped pneumatic spring. *J. Sound Vib.* **1983**, *86*, 191–197.

17. Lee, J.H.; Kim, K.J. Modeling of nonlinear complex stiffness of dual-chamber pneumatic spring for precision vibration isolations. *J. Sound Vib.* **2007**, *301*, 909–926.

18. Jing, X.; Lang, Z. *Frequency Domain Analysis and Design of Nonlinear Systems based on Volterra Series Expansion: A Parametric Characteristic Approach*; Springer: Berlin, Germany; 2015.

19. Xiao, Z.; Jing, X. An SIMO Nonlinear System Approach to Analysis and Design of Vehicle Suspensions, IEEE/ASME trans.*Mechatronics* **2015**, *20*, 3098–3111.

20. Jing, X. Nonlinear characteristic output spectrum for nonlinear analysis and design. *IEEE/ASME Trans. Mechatron.* **2014**, *19*, 171–183.

21. Shin, Y.H.; Kim, K.J. Performance enhancement of pneumatic vibration isolation tables in low frequency range by time delay control. *J. Sound Vib.* **2009**, *321*, 537–553.

22. Moon, J.H.; Lee, B.G. Modeling and sensitivity analysis of a pneumatic vibration isolation system with two air chambers.*Mech. Mach. Theory* **2010**, *45*, 1828–1850.

23. Li, Z.; Li, X.; Chen, X. Generic vibration criteria-based dual-chamber pneumatic spring vibration isolation table design. *Proc. Inst. Mech. Eng. B J. Eng. Manuf.* **2014**, *228*, 1621–1629.

24. Liu, H.; Lee, J.C. Model development and experimental research on an air spring with auxiliary reservoir. *Int. J. Automot. Technol.* **2011**, *12*, 839–847.

25. Zhu, S.; Wang, J.; Zhang, Y. Research on theoretical calculation model for dynamic stiffness of air spring with auxiliary chamber. In Proceedings of the IEEE Vehicle Power and Propulsion Conference, Harbin, China, 3–5 September 2008.

26. Pu, H.; Luo, X.; Chen, X. Modeling and analysis of dual-chamber pneumatic spring with adjustable damping for precision vibration isolation. *J. Sound Vib.* **2011**, *330*, 3578–3590.

27. Erin, C.; Wilson, B.; Zapfe, J. An improved model of a pneumatic vibration isolator: Theory and experiment. *J. Sound Vib.* **1998**,*218*, 81–101.

28. Heertjes, M.; van de Wouw, N. Nonlinear dynamics and control of a pneumatic vibration isolator. *J. Vib. Acoust.* **2006**, *128*, 439–448.

29. Lee, J.H.; Kim, K.J. A method of transmissibility design for dual-chamber pneumatic vibration isolator. *J. Sound Vib.* **2009**, *323*, 67–92.

30. Toyofuku, K.; Yamada, C.; Kagawa, T.; Fujita, T. Study on dynamic characteristic analysis of air spring with auxiliary chamber. *JSAE Rev.* **1999**, *20*, 349–355.

31. Rao, C.R. Factorial experiments derivable from combinatorial arrangements of arrays. *Suppl. J. R. Stat. Soc.* **1947**, *9*, 128–139.

32. Taguchi, G. Performance analysis design. *Int. J. Prod. Res.* **1978**, *16*, 521–530.

33. Azouzi, R.; Guillot, M. On-line prediction of surface finish and dimensional deviation in turning using neural network based sensor fusion. *Int. J. Mach. Tools Manuf.* **1997**, *37*, 1201–1217.

34. Green, P.E.; Krieger, A.M.; Wind, Y. Thirty years of conjoint analysis: Reflections and prospects. *Interfaces* **2001**, *31*, S56–S73.

35. Kutner, M.H.; Nachtsheim, C.; Neter, J. *Applied Linear Regression Models*; McGraw-Hill/Irwin: New York, NY, USA, 2004.

7

CHAPTER

Small Scale Field Experiment on Breaking Wave Pressure on Vertical Breakwaters

Paolo Boccotti, Vincenzo Fiamma, Giuseppe Barbaro, Luigi Collia

DICEAM Department, Natural Ocean Engineering Laboratory, Mediterranea University, Reggio Calabria, Italy

ABSTRACT

A small scale field experiment (SSFE) was performed on vertical breakwaters in the surf zone. The following are some of the findings. Wind seas may yield breaking wave pressure notwithstanding some large deepwater wave steepness, and small elevation of the wall above the mean water level. Caisson breakwaters can withstand some exceptionally high impulsive force peaks (even twice the weight in still water); whereas, with the same sea state and weight, a breakwater composed of layers of solid concrete blocks is destroyed.

Keywords: Impulsive Breaking Wave Pressure, Vertical Breakwater, Small Scale Field Experiment, Breakwater Failure

1. INTRODUCTION

A small scale field experiment (SSFE) is an experiment performed like in a big wave tank, in sea areas or lakes where wind seas have some small sizes [1] . The paper shows the results of a SSFE on vertical breakwaters in the surf zone.

It is well-known since [2] that, under the action of breaking waves, a vertical breakwater may be subject to pressure peaks of high intensity and short duration. These are called impulsive breaking wave pressures (IBWP).

A first item of the SSFE was to test the questionnaire of [3] for the risk of occurrence of IBWP.

A second item was to analyze resistance or failure of breakwaters under breaking waves. Many tests have been made on this subject on wave flumes [4] . A novelty of this paper is that two vertical breakwaters with a different configuration are beaten by the same wind sea. The interest is in the fact that, notwithstanding the configurations of the two breakwaters is similar to each other, one fully collapses and the other perfectly resists. Realizing the reason for this seeming paradox enables us to choose the more effective configuration in the preliminary design of a breakwater.

2. METHOD

The experiment was performed off the shore of the NOEL (Natural Ocean Engineering Laboratory) on the Eastern coast of the Straits of Messina. A vertical breakwater was made with concrete blocks (Figure 1). In a configuration A the blocks were pressed together by some tensioned iron cables so as to obtain a compact structure like that of a caisson breakwater; in a configuration B the blocks were not pressed together. The central section consisted of a small iron caisson being ballasted with pig-iron and equipped with 16 pressure transducers. The incident waves were recorded by an ultrasonic wave probe on deep water. The sample rate was of 100 Hz. Each record had a duration of five-minutes. The average water depth at the base of the upright section was of 0.5 m, and the average depth of the seabed at the breakwater was of 0.6 m (see Figure 1). The bottom slope was 1/10. The tide amplitude was about 0.15 m.

The period T_h of the largest waves of a sea state was obtained from the autocovariance Ψ(T) of the free surface displacement on deep water [1] [5] ; the period $T_{1/3}$ mean of the wave periods of the largest one third waves of a sea state was obtained from the record of the waves on deep water. The average difference $|T_h - T_{1/3}|$ proved to be within the 1.8% of T_h. The narrow bandedness parameter Ψ^* was obtained from Ψ(T) [1] [5] . The wave force F on the central upright section of the breakwater, proceeded from the difference between the actual sectional force obtained from the measurements of the 16 pressure transducers, and the hydrostatic sectional force. The wave pressure pw was obtained as the difference between the actual pressure on the wall and the hydrostatic pressure. The tide level (being necessary to estimate the water depth and hence the hydrostatic pressures and the hydrostatic sectional force) was estimated from the measurement of the free surface displacement made by the ultrasonic probe on deep water.

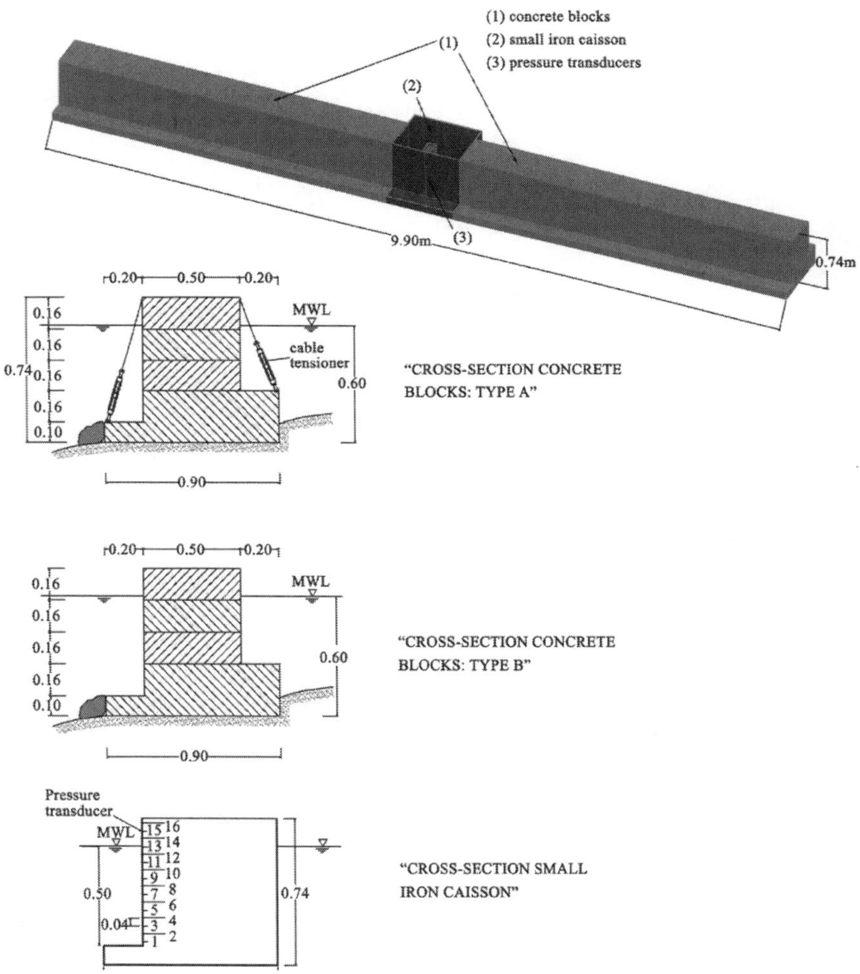

Figure 1. The breakwater of the SSFE. In a stage of the experiment the cross-section had the configuration A. In another stage the cross-section had the configuration B.

3. RESULTS

With breakwater A, we obtained 24 five-minute records of wind seas. These had the following characteristics:

$$0.45\ \mathrm{m} < d < 0.52\ \mathrm{m} \tag{1}$$

$$0.41\ \mathrm{m} < H_{s0} < 0.52\ \mathrm{m} \tag{2}$$

$$0.25 < W/H_{s0} < 0.40 \tag{3}$$

$$0.60 < \psi^{*} < 0.70 \tag{4}$$

$$2.27\ \mathrm{s} < T_{h} < 2.56\ \mathrm{s} \tag{5}$$

$$2.45\ \mathrm{s} < T_{p} < 3.10\ \mathrm{s} \tag{6}$$

$$0.043 < H_{s0}/L_{p0} < 0.056 \tag{7}$$

The spectra on deep water were the classic ones of wind seas (see a few examples on Figure 2). These wind seas were superimposed on a very small swell component (see the small peak of the spectrum on about 0.13 Hz).

The dominant direction of wind seas on deep water off the beach at Reggio Calabria makes an angle within 25° with the orthogonal to the contour lines (this has been confirmed by a lot of small scale field experiments [1] [5]). Hence the angle between the dominant wave direction and the orthogonal to the breakwater was definitely smaller than 25°, because of refraction.

Let us consider the questionnaire [3] , on the danger of IBWP. We are dealing with a case of rubble mound sufficiently small to be considered negligible. The angle between the wave direction and the line normal to the breakwater is less than 20°. The sea bottom slope (1/10) is steeper than 1/50. The breaking point of the progressive wave (in absence of the structure) is located only slightly in front of the breakwater. All of this is proper for the occurrence of IBWP. However there are two conditions which, on the basis of the aforesaid questionnaire, let us conclude that there should be little danger of IBWP; these are

1) a large wave steepness;
2) a small crest elevation that allows much overtopping.

Thus finding IBWP in these wave records was an unexpected outcome of the experiment. This finding is evident in Figure 3, Figure 4 and Figure 5 which show the three largest sectional wave forces F in the 24 records. Each figure shows the pressure distribution $p_w/(\gamma H_{s0})$ at the time instant of the maximum sectional wave force, and F(t). We see that the largest values of p_w exceed 6 (γH_{s0}).

Breakwater A was able to withstand, without sliding, a F_{max} of 13.5 kN/m (see Figure 3) which is twice the P^*! On the opposite, breakwater B after 45 minutes of wind seas very similar to the wind seas that attacked breakwater A, was destroyed (see Figure 6). Only the small iron caisson with the transducers remained stable. In all, the 45 minutes in which breakwater B was destroyed included 9 five-minute records whose characteristics were as follows

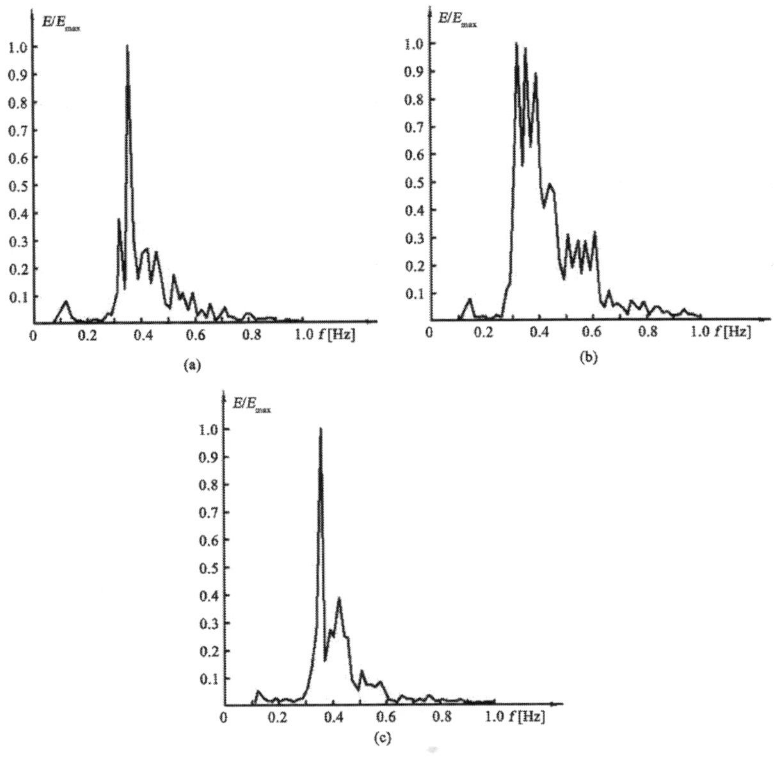

Figure 2. Frequency spectra of deep water waves: (a) rec. 1373; (b) rec. 1424; (c) rec. 1390.

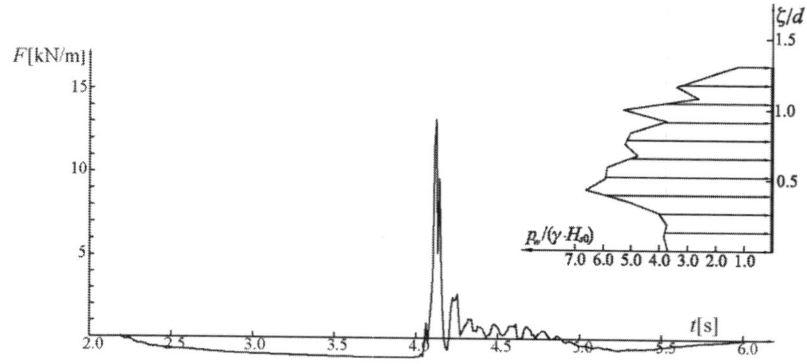

Figure 3. Time history of the largest sectional force, and distribution of the wave pressure yielding the force peak. Breakwater A, rec. 1373.

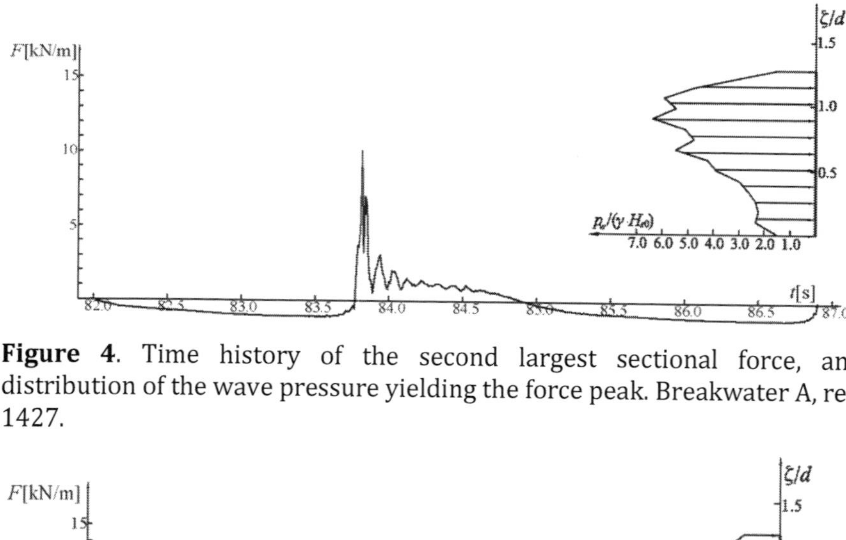

Figure 4. Time history of the second largest sectional force, and distribution of the wave pressure yielding the force peak. Breakwater A, rec. 1427.

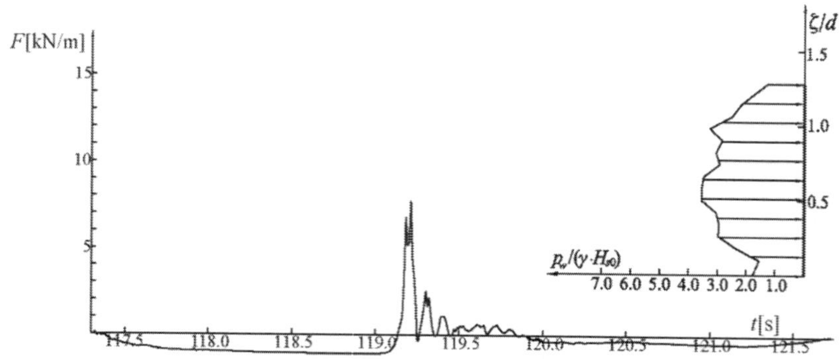

Figure 5. Time history of the third largest sectional force, and distribution of the wave pressure yielding the force peak. Breakwater A, rec. 1390.

Figure 6. Breakwater B destroyed after 45 minutes of wind seas with the characteristics (8 - 14).

$$0.52\,\text{m} < d < 0.53\,\text{m} \quad (8)$$

$$0.40\,\text{m} < H_{s0} < 0.56\,\text{m} \quad (9)$$

$$0.21 < W/H_{s0} < 0.28 \quad (10)$$

$$0.57 < \psi^{*} < 0.71 \quad (11)$$

$$2.36\,\text{s} < T_h < 2.79\,\text{s} \quad (12)$$

$$2.45\,\text{s} < T_p < 3.10\,\text{s} \quad (13)$$

$$0.040 < H_{s0}/L_{p0} < 0.051 \quad (14)$$

As said these characteristics were close to the characteristics (Equations (1)-(7)) of the wind seas that attacked breakwater A.

Figure 7 shows the time history of the largest sectional wave force recorded on breakwater B. Notwithstanding the sea state is very similar to that of the

24 records with configuration A, the wave force proves to be markedly smaller than that recorded with configuration A (compare Figure 7 with Figure 3, Figure 4 and Figure 5). The fact that the vertical wall was being destroyed (many concrete blocks were displaced from their original location) could be a cause for the wave pressure being less intense.

4. ANALYSIS OF RESISTANCE/FAILURE

The horizontal displacement of a section of the breakwater is estimated by numerical integration of the equations:

$$s_{y_{i+1}} = s_{y_i} + 0.5\left(v_{y_i} + v_{y_{i+1}}\right)\Delta t \tag{15}$$

$$v_{y_{i+1}} = \mathrm{Max}\left(0, v_{y_i} + 0.5\left(a_{y_i} + a_{y_{i+1}}\right)\Delta t\right) \tag{16}$$

where $i = 1, 2, 3, \cdots$ denotes a sample instant and a_{y_i} is estimated by means of:

$$a_{y_i} = \left(F_i - \mu R_{V_i} - R_p\right) / \left(M + C_{in}\left(M - P^*/g\right)\right) \tag{17}$$

The breakwater slides whenever a_y gets greater than zero. This occurs when the horizontal wave force F exceeds the resistance of the soil which is the sum of the friction force exerted on the base of the breakwater (μ R_v) and the normal force exerted on the rear wall (R_p).

Let us apply these equations in order to realize how breakwater A was able to withstand the wind seas with a F_{max} twice the P^*. To this end we firstly consider only the action exerted by waves on the upright section (that is we neglect the uplift force). With μ = 0.6, C_{in} = 2.0 and f = 40° (f serves to estimate the passive thrust of the sloping soil) and with the time series data of F(t) of the 24 records, Equations (15)-(17) yield 21 small slides giving a cumulative total displacement of 0.020 m.

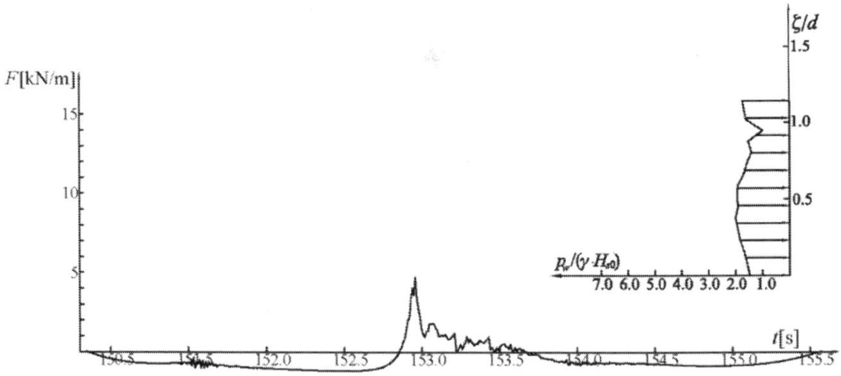

Figure 7. Time history of the largest sectional force (and distribution of the wave pressure yielding the force peak) during the 45 minutes in which breakwater B was broken down by waves.

The fact that the expected displacement was very small notwithstanding the F_{max} was twice the $P^*$is due to the very short duration of the IBWP. The fact that the actual displacement was even smaller (we did not observe any displacement) cannot be attributed to the friction factor: with a m of 0.65 instead of 0.60, the displacement would be reduced only from 0.020 m to 0.014 m; nor it can be attributed to the value of the passive thrust of the sloping soil: on estimating this thrust with a f of 45° instead of 40°, the displacement would be reduced only from 0.020 m to 0.018 m.

The uplift wave force should increase the displacement, in that it is expected to reduce R_v at the time instants of the IBWP on the upright section, even though this effect should not be important. In fact this effect should be much smaller with our foundation (Figure 8(a)) than with a simple foundation like that of Figure 8(b). In conclusion with Equations (15)-(17) we expect a small displacement of breakwater A of the order of a few centimeters, whereas the actual displacement was zero. We believe that this higher resistance of breakwater A is due to the effect of ground which is elastically deformed under an application of an impulsive breaking wave pressure (this topic was approached by [6]). However, the deformation of soil here would require a 3D model because the waves are short-crested; and this is beyond recent practice [4] . Moreover, it would be necessary to measure the wave force on the whole length of the breakwater, and not only on one section.

Equations (15)-(17) may be applied also to predict the displacement of a block on another block of breakwater B (of course in this case R_p must be taken zero). With the time series data of $p_w(t)$ of the 9 five-minute records

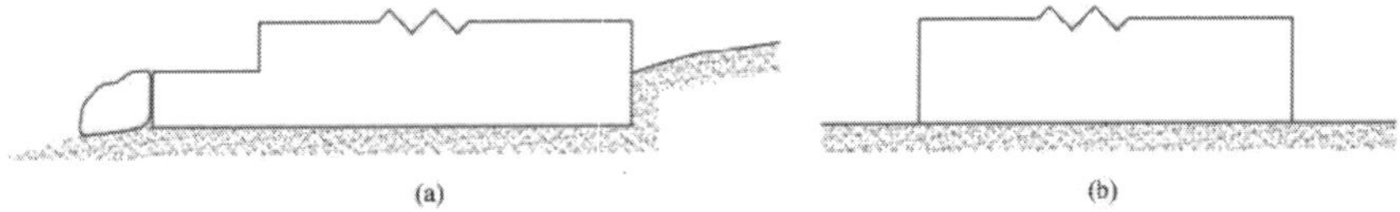

Figure 8. (a) Foundation of breakwaters A or B; (b) a simple foundation considered in our analysis.

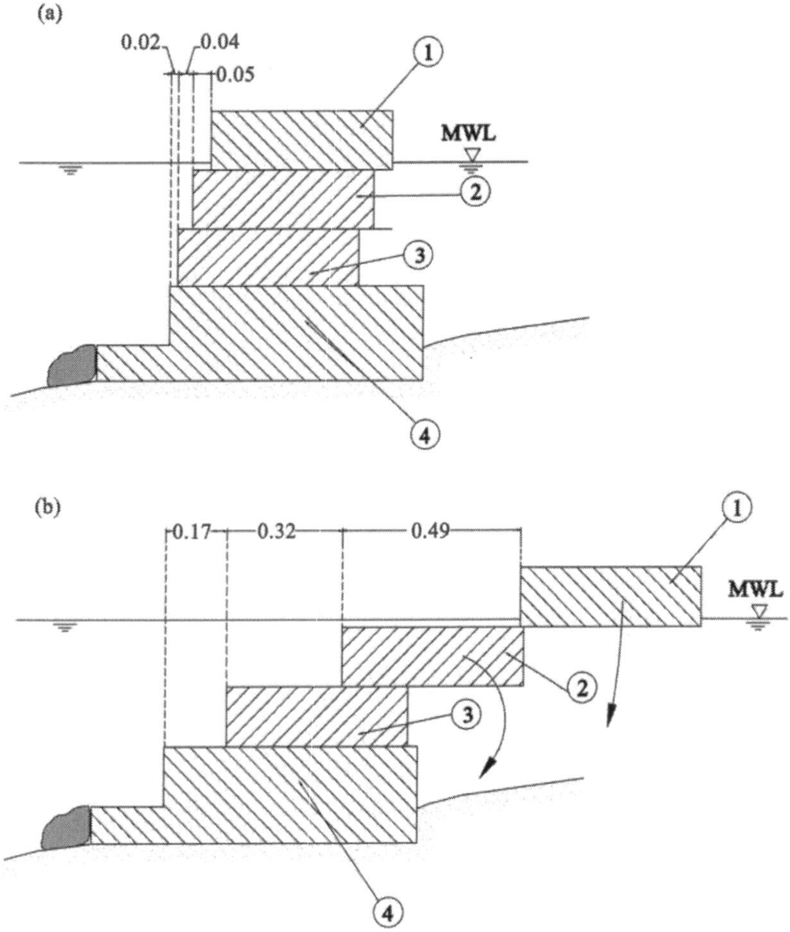

Figure 9. Estimated displacement of the blocks in the 45 minutes in which breakwater B was destroyed. The uplift force at the base of the blocks: (a) is neglected; (b) is taken into account.

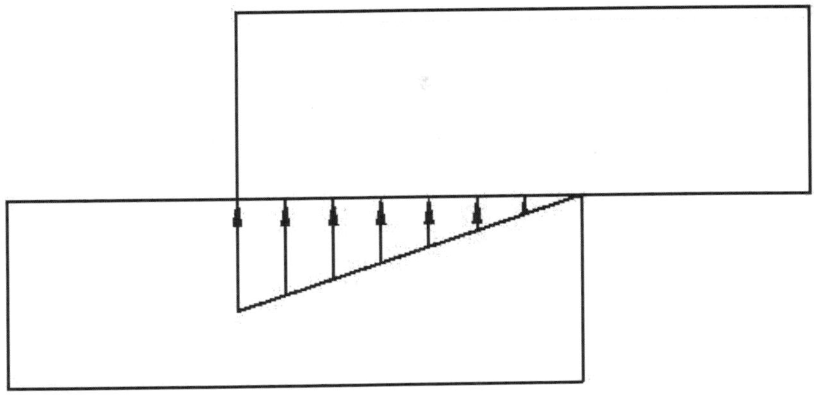

Figure 10. Scheme of the uplift force at the base of a block.

in which breakwater A collapsed we obtain the following result with $\mu = 0.6$, $C_{in} = 2$, and assuming $R_v = P^*$:

1) the displacement of block 1 with respect to block 2 is of 0.05 m as a result of 35 small slides;
2) the displacement of block 2 with respect to block 3 is of 0.04 m as a result of 46 small slides;
3) the displacement of block 3 with respect to block 4 is of 0.02 m as a result of 46 small slides.

These displacements obtained on neglecting the uplift force at the base of each block (R_v assumed to be equal to P^*) are shown in Figure 9(a). The displacements obtained on taking the uplift force into account are shown in Figure 9(b), wherein the situation becomes as follows:

1) the displacement of block 1 with respect to block 2 becomes of 0.49 m as a result of 63 small slides;
2) the displacement of block 2 with respect to block 3 becomes of 0.32 m as a result of 97 small slides;
3) the displacement of block 3 with respect to block 4 becomes of 0.17 m as a result of 86 small slides.

The uplift force is estimated with the usual linear scheme (see Figure 10). It is evident, from a comparison between Figure 9(b) and Figure 9(a), that a crucial role in the collapse of breakwater B was played by the uplift force at the base of each individual block. Of course from Figure 9(b) we may expect that blocks 1 and 2 fall; and this actually occurred.

5. CONCLUSIONS

The questionnaire [3] provides a most clear way to approach the problem of the danger of IBWP. However as Goda himself notes "it is advisable to resort to hydraulic model testing whenever there remains the suspicion of the generation of impulsive breaking wave pressure". According to the questionnaire, if the rubble mound is sufficiently small to be considered negligible, it should not be easy that IBWP are exerted by wind seas attacking nearly orthogonally.

Another reason for which there should have been little danger of IBWP was that the crest elevation of our breakwater was so small as to allow much overtopping: crest elevation between 1/5 and less than 1/2 of H_{s0}.

Nevertheless, we recorded some well evident IBWP on breakwater A.

Because of the fact that:

1) the duration of IBWP was very small;

2) the effect of IBWP on the uplift force was probably negligible, due to the configuration of the foundation;

the displacement of breakwater A with our 24 five-minute records of wind seas should have been within a few centimeters. The fact that no displacement was observed after the action of these wind seas could be explained as an effect of ground which was elastically deformed under an application of an impulsive breaking wave pressure.

Under essentially the same wind seas which were perfectly withstood by breakwater A, breakwater B rapidly collapsed. From our analysis based on Equations (15)-(17) with the time series data of the p_w on the wall, we conclude that the uplift force at the base of each block was the primary cause of the collapse.

As explained in Section 3.7 of [1] , the crucial requirement for a SSFE at some site is "to be reasonably sure that a sea state with some prescribed characteristics will occur after a short waiting time at this site". For this SSFE the prescribed characteristics were: water depth d_n at the breakwater (inclusive of the tide) between 0.55 m and 0.65 m; (d between 0.45 m and 0.55 m); H_{s0} between about 0.40 m and 0.50 m; type of waves: wind seas ($\psi^* > 0.60$); H_{s0}/L_0 covering range Equation (18) of most interest. After 25 years of experiments in the sea of Reggio Calabria we know that we may obtain these conditions. A novelty with respect to the previous SSFEs was in the narrow window for the water depth: the amplitude of this window was smaller than the tide amplitude (this being about 0.15 m). The conditions on d_n, d and H_{s0} served to have a meaningful ratio W/d as well as to have the breaking point of the progressive wave (in absence of the structure) being located only slightly in front of the breakwater.

We obtained many records of sea states with the prescribed characteristics in less than one month. This represents a step forward in the technique of SSFEs, since it is proved that, at least at some sites, these experiments can be performed even on structures of the surf zone.

REFERENCES

1. Boccotti, P. (2015) Wave Mechanics and Wave Loads on Marine Structures. Elsevier, Amsterdam.
2. Bagnold, R.A. (1939) Interim Report on Wave-Pressure Research. Journal of the Institution of Civil Engineers, 12, 202-226.
3. Goda, Y. (2000) Random Seas and Design of Maritime Structures. World Scientific, Singapore.
4. Cuomo, G., Lupoi, G., Shimosako, K. and Takahashi, S. (2011) Dynamic Response of Composite Breakwaters under Breaking and Non-Breaking Wave Attack. Coastal Engineering, 58, 953-969.
5. Boccotti, P. (2000) Wave Mechanics for Ocean Engineering. Elsevier, Amsterdam.
6. Goda, Y. (1994) Dynamic Response of Upright Breakwaters to Impulsive Breaking Wave Forces. Coastal Engineering, 22, 135-158.

8

CHAPTER

Application of Stationary Phase Method to Wind Stress and Breaking Impacts on Ocean Relatively High Waves

Augustin Daika, Theodule Nkoa Nkomom, Cesar Mbane Biouele*

Laboratory of Earth's Atmosphere Physics, Department of Physics, Faculty of Science, University of Yaoundé I, Yaoundé, Cameroon

ABSTRACT

Wind stress impacts on ocean relatively high waves can be perfectly illustrated by a recurrent phenomenon in the Sahara desert. Indeed, on this area where the surface wind can blow without encountering major obstacle out of the sand dunes, these main targets are gradually eroded and displaced by the wind on dozens of meters. This experience highlights the action of wind on granular targets (clusters of sand or water slides) and motivates studies similar to ours, where we want to simulate impact of wind stress and breaking on the spatio-temporal evolution of the envelope of ocean relatively high waves: Impact which can inappropriately deflect the waves on ships, oil platforms or coastal infrastructures. Euler and Navier-Stokes equations allow a mathematical formulation of the gravity wave motion (ocean waves are considered in our work as a system of water particles which are held together by low surface tension) and wind acts on targets through friction forces or stress. Michel Talon stationary phase method is used to numerically solve the equations that model the impact of wind on a stationary Gaussian.

Keywords: Wind Stress; Granular Targets; Low Surface Tension; Stationary Phase Method; Stationary Gaussian

1. INTRODUCTION

Since men have traveled on the oceans, they are impressed by these relatively hostile huge rivers that inspire respect and fear. As evidence of this fear, many legends have always circulated that such stories express the existence of mermaids shipwrecks [1,2], the ghost ships savagely attack ships, or even more recently beliefs about the Bermuda Triangle, where ships disappear inexplicably. Among these legends is that of rogue waves that correspond in many respects to deep water gravity waves [3]. Many accounts of seamen have alluded to walls of water rising for no reason in the middle of the sea and hitting ships with extraordinary violence. These stories were not credible until 1978, when the cargo ship "Munchen" disappeared under mysterious circumstances. This vessel at the forefront of naval technology was heading in the North Atlantic, with no apparent problems until the night of December 12. Given that the weather service recorded no storm that night, it is reasonable to believe that a rogue wave is the only plausible explanation for this shipwreck. In 1980, Philippe Lijour, captain of the tanker "Esso Lanuedoc", demonstrates the existence of rogue waves with a photo as proof. The existence of rogue waves is now universally recognized [4-9], and many images on the extent of damage caused by these monsters of the ocean are available. However, the physical processes responsible for the formation and spread of the phenomenon and its prediction are not completely understood. Contrary to popular belief, Mbane [10] demonstrated that, the natural phenomena like rogue waves are not just spectacular events accessible to routine observations and satellites images. Rogue waves are a combination of complex physical processes that occur under ideal conditions of temperature and humidity [10]. Numerical computations [11-23] offer tremendous opportunities for approaches of physical phenomena for which analytical solutions are at the present stage of development of mathematics, difficult to obtain. Our previous paper [24] based on application of Benjamin-Feir equations to tornadoes' rogues waves modulational instability in oceans is an important tool for acquiring information on the Kinematics and thermodynamics of atmosphere processes which trigger deep rivers gravity waves. In this regard, tornadoes can give birth to ocean gravity waves whose size varies depending mainly on depth of depression and sea surface temperature [24]. In the Sahara desert, sand dunes are gradually eroded and displaced by the wind on dozens of meters. This experience highlights the action of wind on granular targets (clusters of sand or water slides) and motivates studies similar to ours, where we want to simulate impact of wind stress and breaking on the spatio-temporal evolution of the envelope of ocean relatively high waves: Impact which can inappropriately deflect the waves on ships, oil platforms or coastal infrastructures. Euler and Navier-Stokes equations allow a mathematical formulation of the gravity wave motion (ocean waves are considered in our work as a system of water particles which are held together by low surface tension) and wind acts on targets through friction forces or stress. Michel Talon stationary phase method is used to numerically solve the equations that model the impact of wind on a stationary Gaussian.

2. BASIC FORMULATION OF STATIONARY PHASE METHOD

2.1. Additional Assumptions

The general continuity equation for a fluid is:

$$\frac{\partial \rho}{\partial t}+\frac{\partial \rho u}{\partial x}+\frac{\partial \rho v}{\partial y}+\frac{\partial \rho w}{\partial z}=0 \tag{1}$$

This leads to the continuity equation for an incompressible fluid

$$\frac{\partial u}{\partial x}+\frac{\partial v}{\partial y}+\frac{\partial w}{\partial z}=0 \tag{2}$$

The velocity perpendicular to the surface of the water and perpendicular to the impermeable bottom is zero:

$$\boldsymbol{V}\cdot\boldsymbol{n}=0, \quad \text{at } z=-H \text{ or } z=\eta \tag{3}$$

Here $\boldsymbol{n}$ is the surface normal.

When the bottom is parallel to the undisturbed surface this simplifies to

$$w = 0, \quad z = -H \tag{4}$$

and the kinematic boundary condition at the surface

$$\left(u, v; w - \frac{\partial \eta}{\partial t}\right) \cdot \left(-\frac{\partial \eta}{\partial x}, \frac{\partial \eta}{\partial y}, 1\right) = 0,$$

$$z = \eta \Rightarrow w = \frac{\partial \eta}{\partial t} + u\frac{\partial \eta}{\partial x} + v\frac{\partial \eta}{\partial y} \tag{5}$$

Remembering that the surface of the water is allowed to change with time. The last condition comes from the Newtonian force on a moving fluid element

$$\frac{\partial \boldsymbol{V}}{\partial t} + (\boldsymbol{V} \cdot \nabla)\boldsymbol{V} = \nu\left((\nabla \cdot \nabla)\boldsymbol{V}\right) - \frac{1}{\rho}\nabla P + \boldsymbol{g} \tag{6}$$

For an in viscid fluid this simplifies to

$$\frac{\partial \boldsymbol{V}}{\partial t} + (\boldsymbol{V} \cdot \nabla)\boldsymbol{V} = -\frac{1}{\rho}\nabla P + \boldsymbol{g} \tag{7}$$

When the flow is irrotational

$$\nabla \cdot \boldsymbol{V} = 0 \tag{8}$$

We can introduce the velocity potential

$$V = (u, v, w) = \left(\frac{\partial \phi}{\partial x}, \frac{\partial \phi}{\partial y}, \frac{\partial \phi}{\partial z} \right) = \nabla \phi \tag{9}$$

Giving the continuity equation

$$\frac{\partial^2 \phi}{\partial x^2} + \frac{\partial^2 \phi}{\partial y^2} + \frac{\partial^2 \phi}{\partial z^2} = \nabla \phi = 0 \tag{10}$$

The kinematic boundary condition at the bottom

$$\nabla \phi \cdot \boldsymbol{n} = 0, \quad z = -H \tag{11}$$

The kinematic boundary condition at the surface

$$\frac{\partial \phi}{\partial z} = \frac{\partial \eta}{\partial t} + \nabla_{\perp} \phi \cdot \nabla_{\perp} \eta, \quad z = \eta$$

When integrating (7) with respect to x, y, z we get the Bernoulli equation, the arbitrary functions of integration $C_1(y,z,t)$, $C_2(x,z,t)$, $C_3(x,y,t)$ must be the same function C(t), which can be absorbed by the velocity potential yielding exactly the same flow

$$\frac{\partial \phi}{\partial t} + \frac{1}{2}(\nabla \phi)^2 + g\eta = -\frac{P}{\rho}, \quad z = \eta$$

Here we have made the assumption that $\boldsymbol{g}$ is constant $\boldsymbol{g} = (0,0,-g)$ making the gravitational force conservative and making it possible to define a potential energy.

Furthermore we have made the assumption that the surface tension can be neglected. At the surface, z = h, for water flows the space above the water is the atmosphere in where the pressure is almost constant along the surface, as the density of air only about 1/800 times that of water. As this constant pressure has no important influence on the solution, we can put P = 0 giving the dynamical boundary condition at the surface of the water.

$$\frac{\partial\phi}{\partial t}+\frac{1}{2}(\nabla\phi)^2+g\eta=O$$

Equations (10), (11), (12), (14) are basis for all the following calculations. By introducing the stream function $\psi(x,y,t)$, defined by $\psi(x,y,t)=\phi(x,y,\eta,t)$, equations (12) and (14) become:

$$\eta_t+(\nabla\eta).(\nabla_\perp\psi)-\phi_z\left(1+(\nabla_\perp\eta)^2\right)=0 \tag{15}$$

$$\psi_t+g\eta+\frac{1}{2}(\nabla_\perp\psi)^2-\frac{1}{2}\phi^2\left(1+(\nabla_\perp\eta)^2\right)=0 \tag{16}$$

2.2. Stationary Phase Formulation of Euler Equations

The spatio-temporal evolution of the surface elevation $\eta(x,y,t)$ can be described by the reduced equation (17) of a linear wave

$$i\frac{\partial a}{\partial X}=\frac{\partial^2 a}{\partial T^2} \tag{17}$$

The general solution of the surface elevation can be expressed as a Fourier integral

$$a(X,T)=\int_{-\infty}^{+\infty}a(\Omega)\cdot\exp\left[i(k\cdot X-\Omega T)\right]\mathrm{d}\Omega \tag{18}$$

where $a(\Omega)$ is obtained from (19) and

$$a(X=0,T)=\int_{-\infty}^{+\infty}a(\Omega)\cdot\exp i\cdot\Omega\cdot T\mathrm{d}\Omega \tag{19}$$

$$a_X\left(X=0,T\right)=\int_{-\infty}^{+\infty} ik\cdot a\left(\Omega\right)\cdot \exp i\cdot\Omega\cdot T d\Omega \quad (20)$$

Boundary conditions below are used: $a_X\left(0,T\right)=0$; $a\left(0,T\right)=a_0\left(T\right)$. Fourier inverse transformation gives:

$$a\left(\Omega\right)=\frac{1}{2\pi}\int_{-\infty}^{+\infty} a_0\left(T\right)\cdot \exp\left(-i\cdot\Omega\cdot T\right) dT \quad (21)$$

Which solution is

$$a\left(X,T\right)=\frac{1}{2\pi}\int_{-\infty}^{+\infty}\int_{-\infty}^{+\infty} a_0\left(T\right)\exp\left(-i\Omega T\right)\cdot\exp\left[i\left(k\cdot X-\Omega T\right)\right] d\Omega dT$$

Using stationary phase method [25,26], we obtain the surface elevation $\eta\left(x,y,t\right)$ expression

$$\eta\left(x,t\right)=\frac{A_0}{2\left(1+16\Omega^4\cdot\varepsilon^2\cdot k_0^2\cdot x^2\right)^{\frac{1}{4}}}\times\exp\left(\frac{-\Omega^2\cdot w_0^2\left(t-x/c_g\right)^2}{1+16\Omega^4\cdot\varepsilon^2\cdot k_0^2\cdot x^2}\right)$$
$$\times\cos\left(-\frac{\arctan\left(4\Omega^2\cdot\varepsilon\cdot k_0\cdot x\right)}{2}-\frac{4\Omega^2\cdot w_0^2\left(t-x/c_g\right)^2\cdot\varepsilon\cdot k_0\cdot x}{1+16\Omega^4\cdot\varepsilon^2\cdot k_0^2\cdot x^2}\right)$$

where e corresponds to the wave steepness; k_0, the wave number;

$$\left|a\right|_{\max}=\frac{a_0}{\left[1+16\Omega^4 x^2\right]^{1/4}}\ ;\quad \Omega\left(t,x\right)=\frac{\partial \arg\left(a\right)}{\partial t}=-\frac{8\Omega^4 t\cdot x}{1+16\Omega^4 x^2}$$

3. RESULTS AND DISCUSSIONS

Figures presented in this manuscript provide evidence of winds stress and breaking barrier impacts on Gaussian waves. Figure 1 shows the Gaussian primary profile of the wave which is used as the target throughout the simulations. Amplitude A_0 and Gaussian steepness will undergo deformations that depend on the magnitude of the wind and the proximity of the barrier (wave-barrier) located 20 m from x coordinate origin. On Figure 2(a), colors variation gives a precise idea of the impact of winds stress on the Gaussian. One can immediately notice the existence of an intense activity of wind (dark red) in the region bounded by x and t values listed in quotes {-8< x < -6; t >0}. Otherwise, the wind attacks the Gaussian basis like it was trying to move the entire structure parallel to its direction. It should however be noted that water particles located in the center of the Gaussian suffer no winds' influence (due to the dark blue color at x = 0), the entire movement of Gaussian is then excluded and the only way allowed to the upstream side particles is to climb the Gaussian and then due to gravity fall on downstream side as illustrated on Figure 2(b). Hence the dark red color of the region bounded by {3 □ x□ 8; t □ 0} which predicts (as t < 0) activity on the downstream side of the Gaussian.

Figures 3(a)-(d) describe the Amplification Factor (A/A_0) as a function of wind-magnitude. Amplification Factors are listed in Table 1. According to this table, Amplification Factor increases exponentially. All these simulations give an indication of the height to impose a barrier to prevent it being crossed by rogue waves.

Figures 4(a)-(d), give a 3D representation of surface elevation $\eta(x,t)$ as a function of space and time.

One can see exactly how the target is pushed towards the barrier as the green color representing a calm ocean surface, gradually gains space when going from Figures 4(a)-(d).

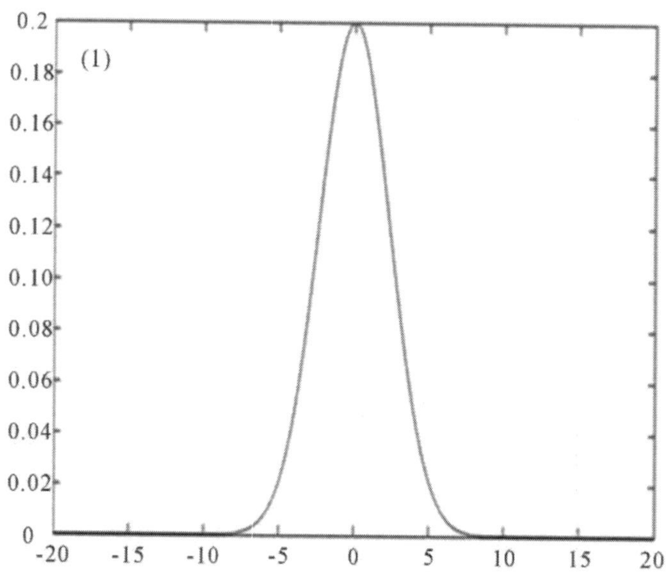

Figure 1. Profile of the primarily wave (Gaussian).

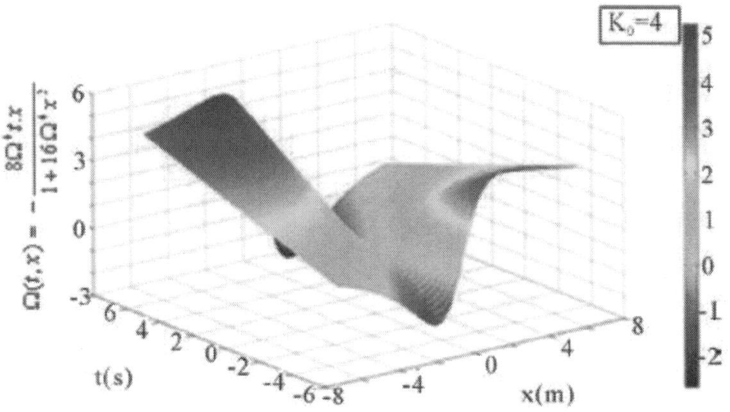

(a)

(b)

Figure 2. (a) Magnitude of winds stress impacts on the target (k_0 = 4 and å = 0.4): k_0 =fundamental wave number; å = small parameter of nonlinearity; (b) Rogue waves' downstream steepness.

Table 1. Amplification factor as function of wind magnitude.

U_{wind}(m/s)	Amplification Factor (dB)
14	0
20	1
30	250
50	9×10^9

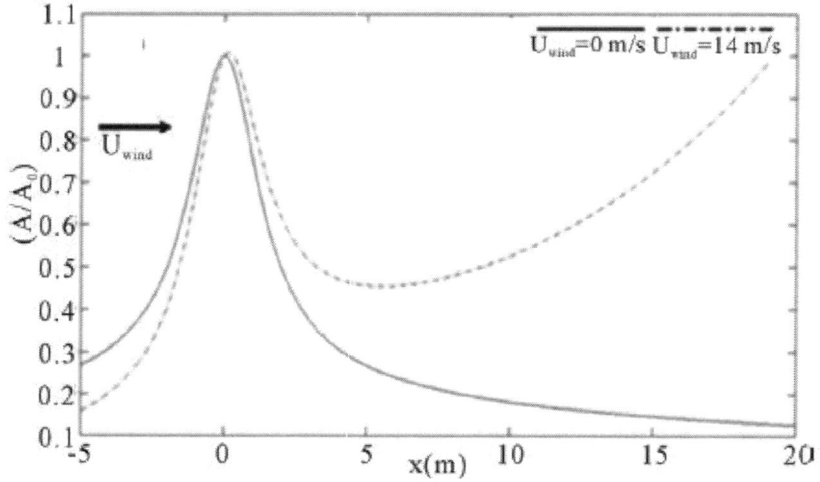
1.1
1
0.9
0.8
0.7
0.6
0.5
0.4
0.3
0.2
0.1
(A/A0)
Uwind
Uwind=0 m/s Uwind=14 m/s
-5
0
5
x(m)
10
15
20

(a)

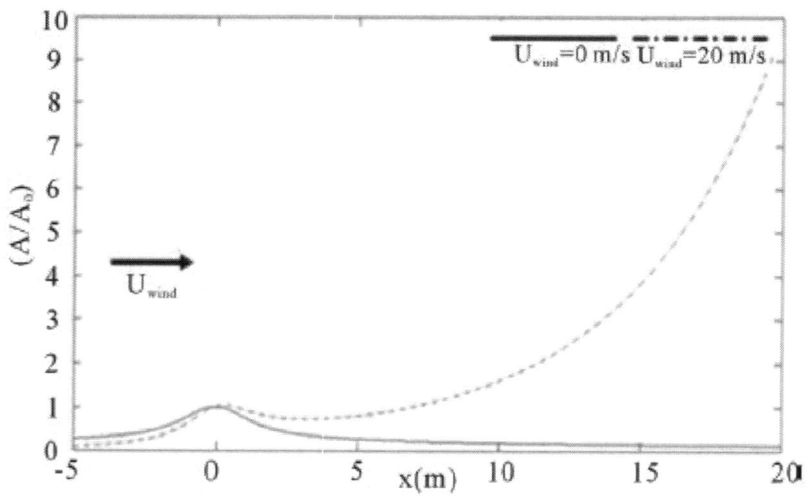
10
9
8
7
6
5
4
3
2
1
0
(A/A0)
Uwind
Uwind=0 m/s Uwind=20 m/s
-5
0
5
x(m)
10
15
20

(b)

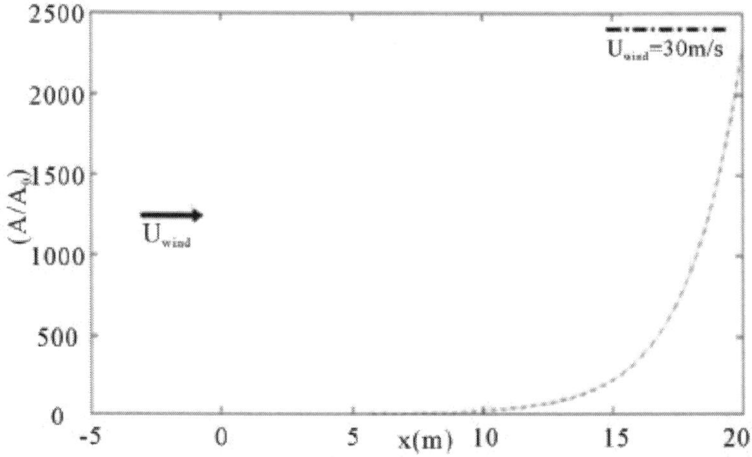

(c)

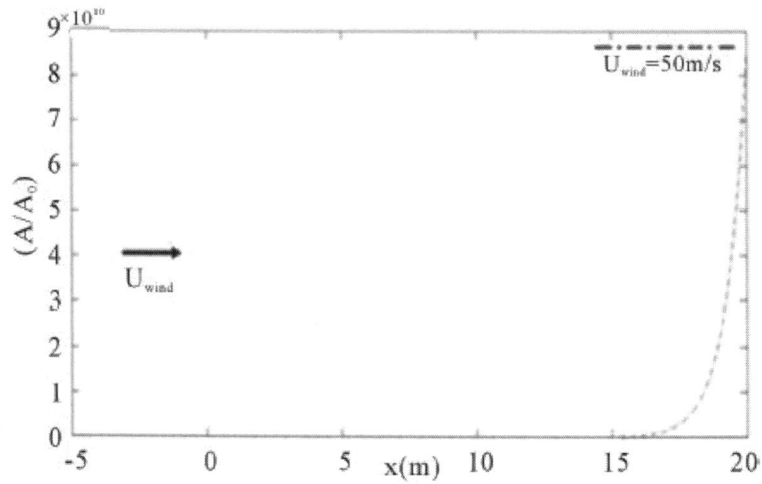

(d)

Figure 3. (a)-(d) Amplification factor for winds stress and breaking barrier impacts on a Gaussian as function of wind magnitude.

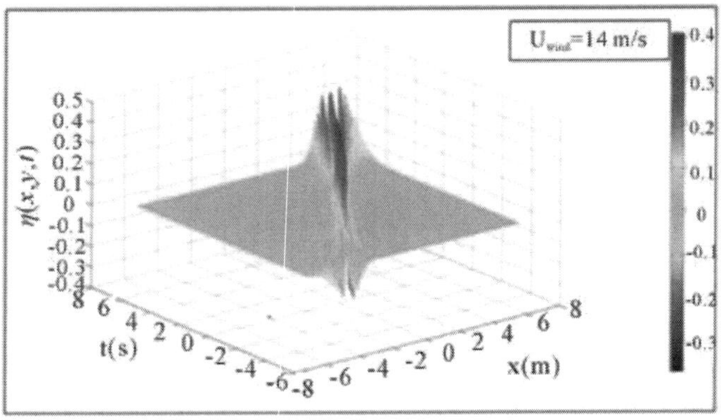
U_wind=14 m/s
η(x,y,t)
t(s)
x(m)

(a)

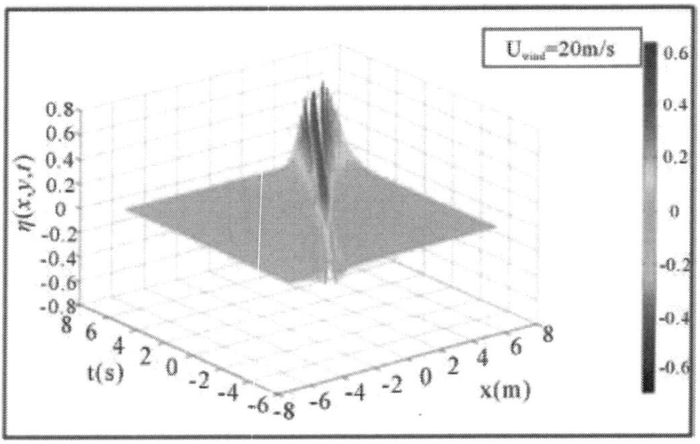
U_wind=20m/s
η(x,y,t)
t(s)
x(m)

(b)

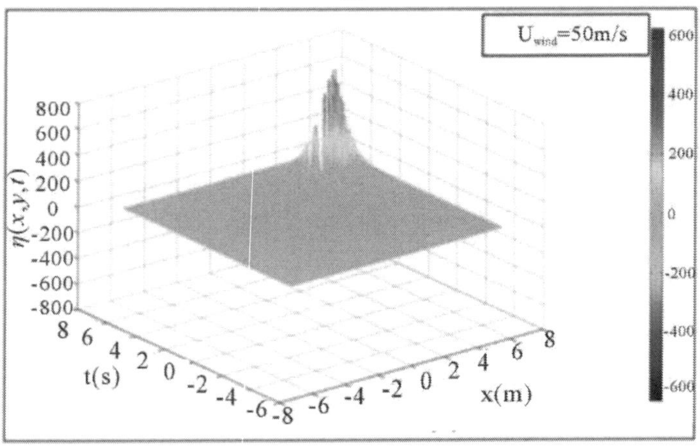
U_wind=50m/s
η(x,y,t)
t(s)
x(m)

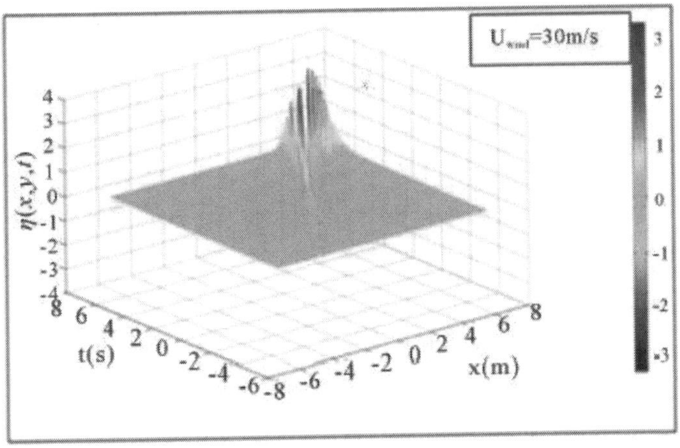

(c)

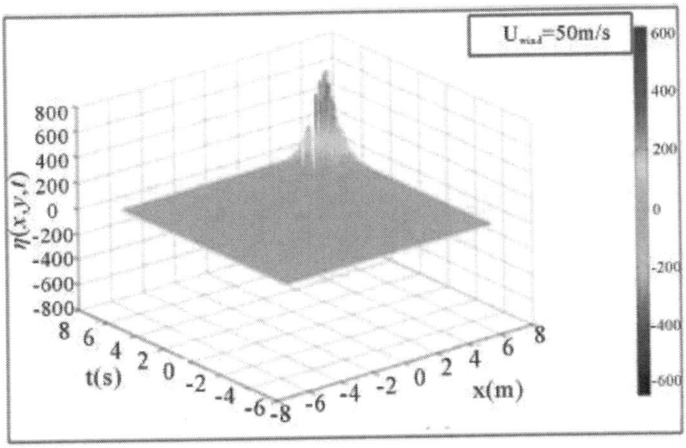

(d)

Figure 4. (a)-(d) Impacts of winds stress and breaking barrier on Surface elevation (3D-representations).

4. CONCLUSION

Winds, as confirmed by the results obtained in this work are a serious threat to the activities on the oceans and shorelines. The wind moves the waves they meet on their way and creates relatively high waves when it causes severe depression over the oceans (e.g., tornadoes or cyclones rogue waves). These results have improved our understanding of the way in which the wind acts on the granular targets and showed how difficult it is to build fences to guard against rogue waves. The choice of inputs of our model (K_0 = 4, e = 0.4) is crucial for the results. But any errors can be corrected by adjusting these inputs progressively as laboratory experiments (or scientific observations) require that, if any. The oceans hostility is associated with both oceans' impressive size that made lose all sense of direction (hence the need to embark compass and GPS) and the fear of unpredictable winds that trigger rogue waves.

REFERENCES

1. J. Touboul, C. Kharif, E. Pelinovsky and J. P. Giovanangeli, "On the Interaction of Wind and Steep Gravity Wave Groups Using Miles' and Jeffreys' Mechanisms," Nonlinear Processes in Geophysics, Vol. 15, 2008, pp. 1023- 1031.
2. T. B. Benjamin, "Instability of Periodic Wave Trains in Nonlinear Dispersive Systems," Proceedings of the Royal Society of London, Vol. 299, No. 1456, 1967, pp. 59-75.
3. T. Karsten and K. Igor, "On Weakly Nonlinear Modulation of Waves on Deep Water," Physics of Fluids, Vol. 12, No. 10, 2000, pp. 24-32.
4. M. Onorato, A. R. Osborne, M. Serio, L. Cavaleri and T. C. Stanberg, "Observation of Strongly Non-Gaussian Statistics for Random Sea Surface Gravity Waves in Wave Flume," Physical Review E, Vol. 70, No. 6, 2004, Article ID: 067302.
5. H. Socquet, A. Juglar, K. Dysthge, K. Trulsen, H. E. Krogstad and J. Liu, "Probability Distributions of Surface Waves during Spectral Change," Journal of Fluid Mechanics, Vol. 000, 2005, pp. 1-21.
6. A. L. Dyachenko and V. E. Zakharov, "Modulation Instability of Stokes Wave-Freak Wave," Journal of Experimental and Theoretical Physics, Vol. 81, No. 6, 2005, pp. 318-322.
7. C. Kharif and E. Pelinovsky, "Physical Mechanism of the Rogue Wave Phenomenon," European Journal of Mechanics/B-Fluid, Vol. 22, No. 6, 2003, pp. 603-634.
8. C. H. Wu and A. Yao, "Laboratory Measurements of Limiting Freak Waves on Current," Journal of Geophysical Research, Vol. 109, 2004, pp. 1-18.

9. B. S. White and B. Fomberg, "On the Chance of Freak Waves at Sea," Journal of Fluid Mechanics, Vol. 355, 1998, pp. 113-138.

10. C. M. Biouele, "Hurricanes and Cyclones Kinematics and Thermodynamics Based on Clausius-Clapeyron Relation Derived in 1832," International Journal of Physical Sciences, Vol. 8, No. 23, 2013, pp. 1284-1290.

11. V. E. Zakharov and N. G. Kharitonov, "Instability of Periodic Waves of Finite Amplitude on the Surface of a Deep Fluid," Journal of Applied Mechanics and Technical Physics, Vol. 11, 1970, pp. 747-751.

12. L. Shener, "On Benjamin-Feir Instability and Evolution of Nonlinear Wave with Finite-Amplitude Side Bands," Natural Hazards and Earth System Sciences, Vol. 10, 2010, pp. 2421-2427.

13. S. Leblanc, "Wind-Forced Modulations of Finite—Dephgravity Waves," Physics of Fluids, Vol. 20, No. 11, 2008, Article ID: 116603.

14. K. Batra, R. P. Sharma and A. D. Verga, "Stability Analysis on Nonlinear Evolution Paterns of Modulational Zakharov Equations," Journal of Plasma Physics, Vol. 72, No. 5, 2006, pp. 671-686.

15. M. I. Banner and J. B. Song, "On Determining the Onset and Strength of Breaking for Deep Water Waves. Part ii: Influence of Wind Forcing and Surface Shear," Journal of Physical Oceanography, Vol. 32, No. 9, 2002, pp. 2559- 2570.

16. M. G. Brown and A. Jensen, "Experiments on Focusing Unidirectional Water Waves," Journal of Geophysical Research, Vol. 106, No. C8, 2001, pp. 16917-16928.

17. H. Jeffreys, "On the Formation of Wave by Wind," Proceedings of the Royal Society A, Vol. 107, No. 742, 1925, pp. 189-206.

18. O. M. Philips, "On the Interaction of Waves by Turbulent Wind," Journal of Fluid Mechanics, Vol. 2 1957, pp. 417- 455.

19. V. K. Makin, H. Branger, W. L. Peirson and J. P. Giovanangeli, "Stress above Wind-plus-Paddle Waves: Modeling of a Laboratory Experiment," Journal of Physical Oceanography, Vol. 37, No. 12, 2007, pp. 2824-2837.

20. J. W. Miles, "On the Generation of Surface Wave by Shear Flow," Journal of Fluid Mechanics, Vol. 3, No. 2, 1957, pp. 185-204.

21. L. Shemer, K. Goulitski and E. Kit, "Evolution of Wide Spectrum Unidirectional Wave Groups in a Tank: An Experimental and

Numerical Study," European Journal of Mechanics—B/Fluids, Vol. 26, No. 2, 2007, pp. 193- 219.

22. J. B. Song and M. I. Banner, "On Determining the Onset and Strength of Breaking for Deep Water Waves, Part i: Unforced Irrotational Wave Groups," Journal of Physical Oceanography, Vol. 32, No. 9, pp. 2541-2558.

23. T. Stanton, D. Marshall and R. Houghton, "The Growth of Waves on Water Due to the Action of the Wind," Proceedings of the Royal Society A, Vol. 137, No. 832, 1932, pp. 283-293.

24. A. Daika, H. M. Etoundi, C. M. Ngabireng and C. Mbané- Biouélé, "Application of Benjamin-Feir Equations to Tornadoes' Rogue Waves Modulational Instability Inoceans," International Journal of Physical Sciences, Vol. 7, No. 46, 2012, pp. 6053-6061.

25. M. Talon, "Ondes de Surface," LIPTHE Paris VI-CNRS, 2006.

26. J. W. Miles, "Surface-Wave Generation: A Viscoelastic Model," Journal of Fluid Mechanics, Vol. 322, 1996, pp. 131-145.

9

CHAPTER

Design Considerations for Pile Groups Supporting Marine Structures with Respect to Scour

Yasser E. Mostafa

Department of Irrigation and Hydraulics, Faculty of Engineering, Ain Shams University, Cairo, Egypt

ABSTRACT

Piles supporting marine structures such as jetties, relieving platforms, quay walls and fixed offshore structures are subjected to lateral loads due to berthing and mooring forces, wind, waves, storm surges and current forces. This paper presents some factors that affect the design of pile groups supporting marine structures founded in cohesionless soils. Some main aspects that should be considered in the pile group design are addressed such as pile batter angle, pile group arrangement, pile spacing, pile slenderness ratio and magnitude of lateral static loading. Numerical analyses were conducted to investigate these design aspects with and without impact of scour. Different scour depths were considered to cover the possible root causes of scour around pile groups such as waves, current and ship propeller jets. The study revealed that scour has greater impact on lateral loading of pile groups compared to its impact on single piles. Pile groups with side-by-side arrangement exposed to scour are more critical than single piles and piles groups with tandem arrangement due to the combined effect of scour and pile-soil-pile interaction. It is also concluded that scour protection is not always required. More attention and considerations should be given to scour protection around piles especially if the piles are closely spaced, arranged side-by-side and if slenderness ratio is less than 12.5.

Keywords: Piles; Pile Groups; Marine Structures; Scour; Battered Piles

1. INTRODUCTION

Marine structures such as jetties, seawalls, relieving platforms, quay walls and fixed offshore (jacket type) structures are often supported on pile groups. The foundation piles usually comprise a large portion of the marine structure cost. These piles are usually subjected to large lateral loads induced from waves, currents, vessel berthing and mooring forces. Also, these piles are subjected to scour due to waves, current and ship propeller jets.

Marine structures and bridge piers supported on pile groups can fail due to severe scour. Numerous publications are found in the literature for investigating the scour around piles for bridge piers and a smaller number of publications investigating the scour around marine structures. Moreover, a very limited number of publications regarding the effect of scour on the behavior of pile groups is found in the literature. Vertical pile capacity is composed of friction along pile length and end bearing at pile toe while pile lateral capacity highly depends on the soil conditions surrounding the top one third of pile length. Therefore, scour impact on lateral pile capacity is more significant than the scour impact on vertical pile capacity.

Global scour refers to a general lowering of the ground surface over a wide area. **Figure 1** shows global scour around a fixed offshore structure supported on piles.

Scour around piles varies due to the root cause. For single piles, scour depth in sandy soils (d_s) is 1.3 times pile diameter (d) with a mean of 0.7 [1]. In another meaning, the ultimate scour depth is about 2 times the pile diameter (i.e., $d_s/d = 2$). However, this value can be different for scour due to waves only or due to ship propeller jets only. Some research has been performed to examine the scour around pile groups due to waves and currents such as Sumer and Fredsøe [2], Mostafa and Agamy [3]. However, very limited research has examined how to eva luate the effects of scour on the behavior of pile groups. Among this limited research, recent research focused on piles supporting bridge piers such as Lin et al. [4].

Effect of scour on lateral loading of single piles has been investigated in a few recent publications such as Kishore et al. [5], Lin et al. [6], Mostafa [7] and Ni et al. [8]. Mostafa [7] reported that scour has significant impact on single piles installed in sand and less significant impact on piles installed in clay. Also, Mostafa [7] reported that ultimate lateral capacity for single pile subjected to global scour is found to be about 50% to 70% of ultimate lateral capacity of pile subjected to local scour depending on the scour hole dimension.

Figure 1. Global scour: wide depression around a jacket structure (after Whitehouse [10]).

As piles are usually installed in groups, it becomes necessary to study the effect of scour on the behavior of pile group not just single piles. The combination of scour and pile-soil-pile interaction (i.e., group effect) can lead to a significant reduction in lateral pile capacity and consequently may lead to the failure of marine structures.

This paper presents the impact of global scour around batter and vertical pile groups installed in medium dense sand. The software program GROUP V.7.0 [9] was used in the analyses. For calibration, results from the software were compared with experimental tests on batter pile groups found in the literature.

Different design parameters were investigated in this paper such as pile batter angle, pile group arrangement, pile spacing and pile slenderness ratio. The impact of scour depth on all these parameters was investigated. Only global scour was considered in this study as it has more significant impacts compared to local scour.

Based on the results of the numerical analyses, this paper also provides general recommendations and guidelines on the necessity of using scour protection. Scour protection using riprap or geotextile may be necessary sometimes and may be a waste of money in other cases. The decision to protect the piles from scour depends on the maximum anticipated scour depth based on the root cause of scour and also depends on the pile, soil and loading conditions.

2. NUMERICAL ANALYSIS AND VALIDATION WITH PREVIOUS EXPERIMENTS

In this paper, numerical analyses were conducted to investigate several parameters affecting the design of pile groups supporting marine structures subjected to scour. The software program Group [9] was used in the analysis. GROUP is a 3D software program for analyzing pile groups subjected to axial and lateral loads. A solution requires iteration to accommodate the nonlinear response of each pile in the group model. The program GROUP solves the nonlinear response of each pile under combined loadings. For closely-spaced piles, the pile-soilpile interaction is taken into account by introducing reduction factors for the p-y curves used for each single pile. These reduction factors or called "p-multipliers" are generated based on results of laboratory and field experiments published in the literature.

A comparison between the results from the computer program Group [9], results from the program Piglet [11] and experimental tests of batter pile groups performed by Zhang et al. [12] was conducted. Batter piles are widely used to support marine structures especially for structures subjected to relatively large lateral loads. Zhang et al. [12] carried out 18 different lateral load tests in the centrifuge on 3 × 3 and 4 × 4 fixed head battered pile groups to investigate the effects of vertical load on the group lateral resistance in cohesionless soils. Zhang et al. [12] proved that designs based on standard lateral load tests with small vertical dead loads would be on the safe side.

Figure 2 shows a sketch of the prototype 3 × 3 battered pile group simulated by the centrifuge models. Two pile arrangements were simulated for the 3 × 3 pile group (Zhang et al. [12]). In the first arrangement, the side pile rows were battered forward at 1:8 slope and the middle row was battered reverse at 1:4 slope (referred to as the 6F3R arrangement). In the second arrangement, the side pile rows were battered reverse at 1:8 slope and the middle row was battered at 1:4 slope (3F6R arrangement). The piles were square aluminum with 304 mm in length and 9.5 mm in width. In prototype scales, the width, total length and embedded length of the piles were 0.43, 13.7 and 10.4 m, respectively and the free length from the ground surface to the point of lateral load application was 2.7 m. The piles were three-diameter spaced and rigidly attached to the pile cap. The soil comprised medium dense sand with relative density (D_r) of 55%.

Figure 3(a) shows the load-deflection curve for the 3 × 3 battered piles with 3F6R arrangement. Results from the computer programs Group and Piglet provide reasonable estimate in comparison with the centrifuge experiments. For low lateral loads, the computer programs tends to slightly overestimate the lateral deflection com pared to experimental results; while for the large lateral loads, the computer programs tend to underestimate the lateral deflections. **Figure 3**(b) indicates that, for the same lateral load, pile group with 6F3R arrangement has less deflection than that for the 3F6R arrangement. This can be attributed to the fact that in the case of 6F3R arrangement six piles have steeper batter angle slope (1:4) and three piles have milder batter angle slope (1:8) which provides higher lateral resistance than the case of 3F6R arrangement.

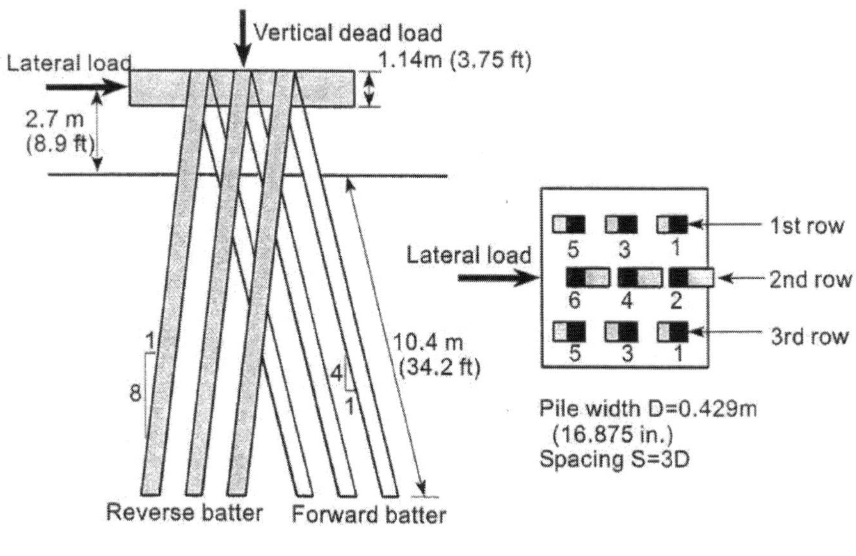

Figure 2. Prototype 3 × 3 battered pile groups tests (after Zhang et al. [12]).

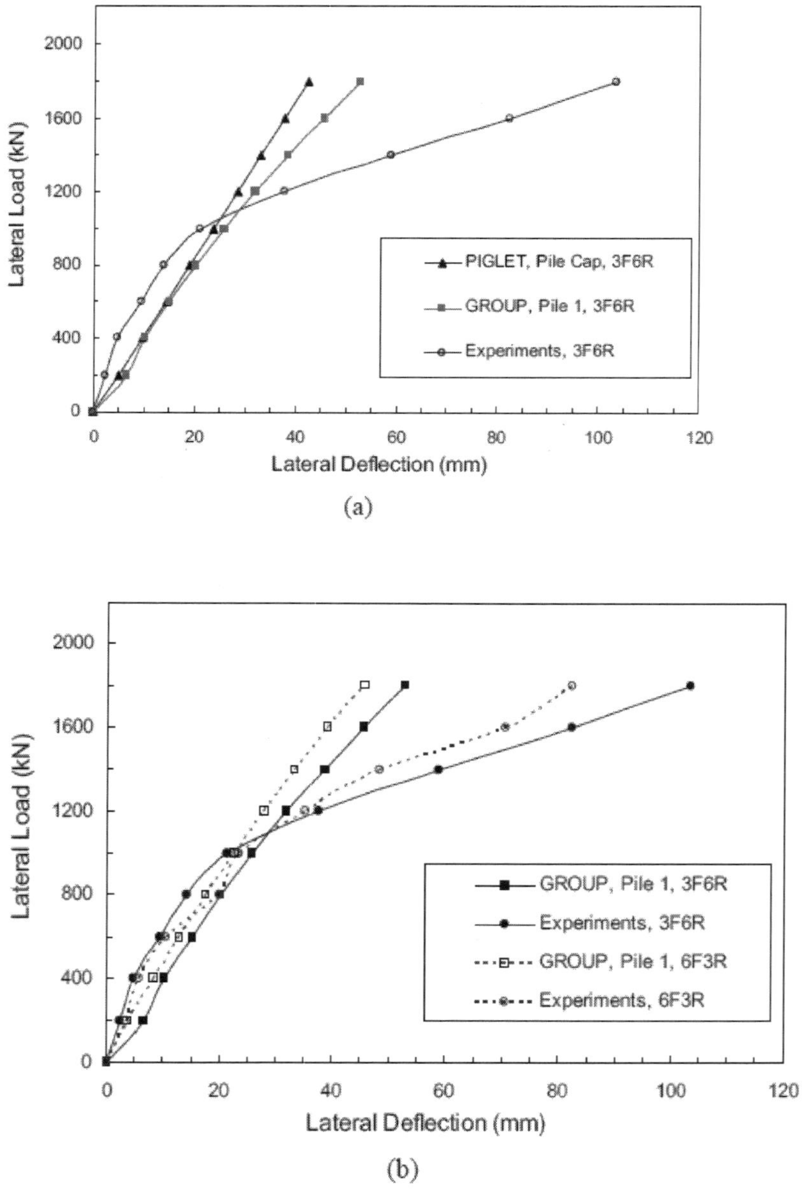

Figure 3. Load-deflection curves for the 3 × 3 battered pile groups considering two pile arrangements: (a) 3F6R arrangement and (b) 6F3R arrangement.

3. EFFECT OF BATTER ANGLE

Batter (raked) piles are widely used to support marine structures as they are effective in restricting horizontal displacement of the foundation subjected to large lateral loads or in a liquefied soil flow environment. The disadvantages of such piles are the larger axial forces in the piles [13]. The forces on piles supporting coastal structures are axial vertical loads due to own weight, loads from trucks and cranes and lateral loads from ship impacts, wave loads and mooring forces. The use of batter piles along with vertical piles increases the overall pile group efficiency. As reported by Rajashree and Sitharam [14], Batter piles are classified as positive batter (slip surface deflects upward) and negative batter (slip surface deflects downward) depending on the formation of slip surfaces [15].**Figure 4** illustrates the negative and positive batter piles.

In this paper, the effect of batter angle was investigated. In the analysis, all piles were assumed to be steel piles with outside diameter (d) of 0.5 m, total pile length (L) of 30 m and cross section area (A) of 0.01944 m^2. The pile spacing to diameter ratio (S/d) was taken to be 3. The distance between point of lateral load at the jetty deck and the seabed (i.e., pile free length, L_f) was assumed to be 5 m. This free length is required for vessels berthing and it varies depending on the vessel draft, under keel clearance and distance between sea level and jetty deck level. All pile heads were assumed to be fixed to the jetty deck. The soil was assumed to be medium dense sand with a unit weight (g) of 17 kN/m^3 and angle of internal friction (f) of 35°. The ratio between pile and soil modulus of elasticity (E_p/E_s) was assumed to be 2000. In the analysis, two piles were considered; one pile has negative batter and the other one has positive batter. Different batter angles of 0°, 5°, 10° and 20° were investigated.

Load-displacement curves for different pile batter angles are shown in **Figure 5**. For the same lateral loading on pile group, a small batter angle significantly reduces the lateral pile displacement. For example, for a lateral load of 400 kN applied to the group at the deck level, the normalized lateral pile displacement (y/d) is 0.23 for vertical piles (b = 0°) and about 0.1 for batter piles with angle b of 5° (i.e., reduction of 57% in pile head displacement). It is also noted that increasing the batter angle from b = 0° to b = 5° increases the lateral capacity from 400 kN to 700 kN (i.e., increase of 75% of lateral capacity).

Figure 6 shows the lateral displacement and bending moment along pile length if the pile group is subjected to lateral load of 100 kN applied at the pile cap. A slight increase in the batter angle significantly reduces the lateral pile displacement and bending moment. It is worth

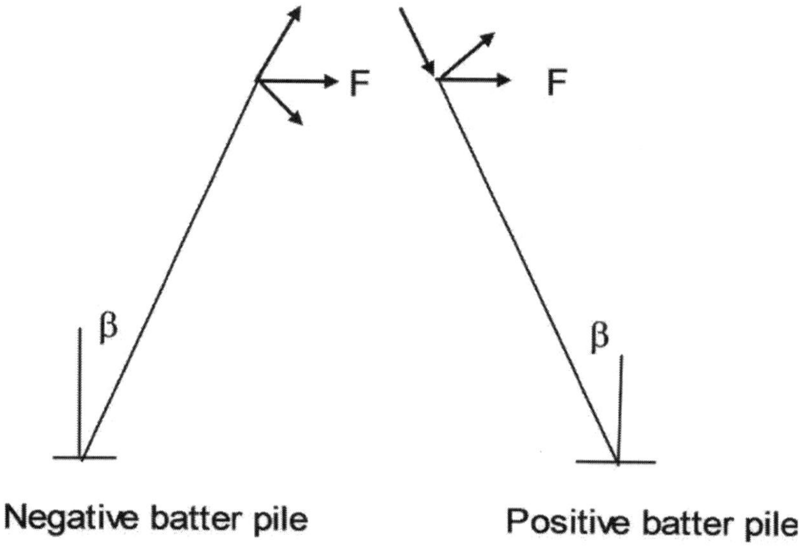

Figure 4. Sketch of negative and positive batter piles.

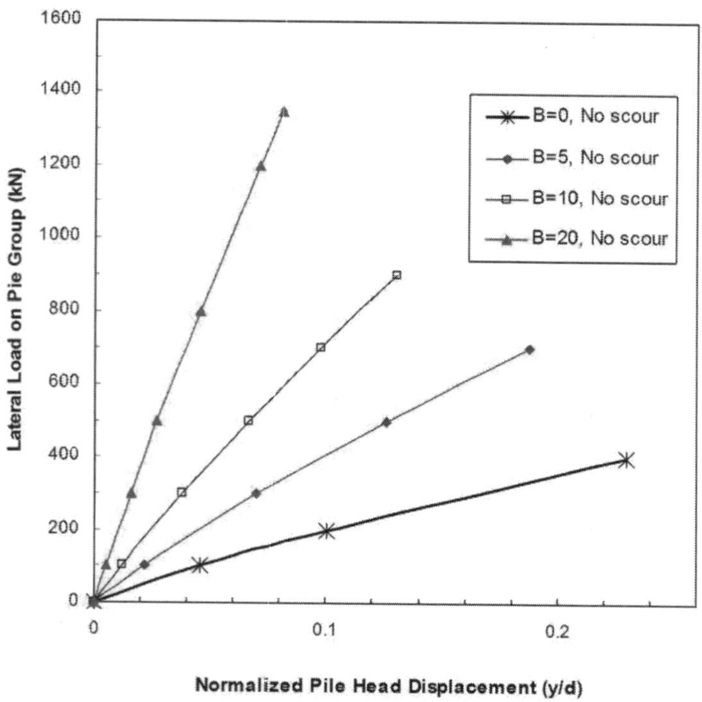

Figure 5. Load-displacement curves for different pile batter angle.

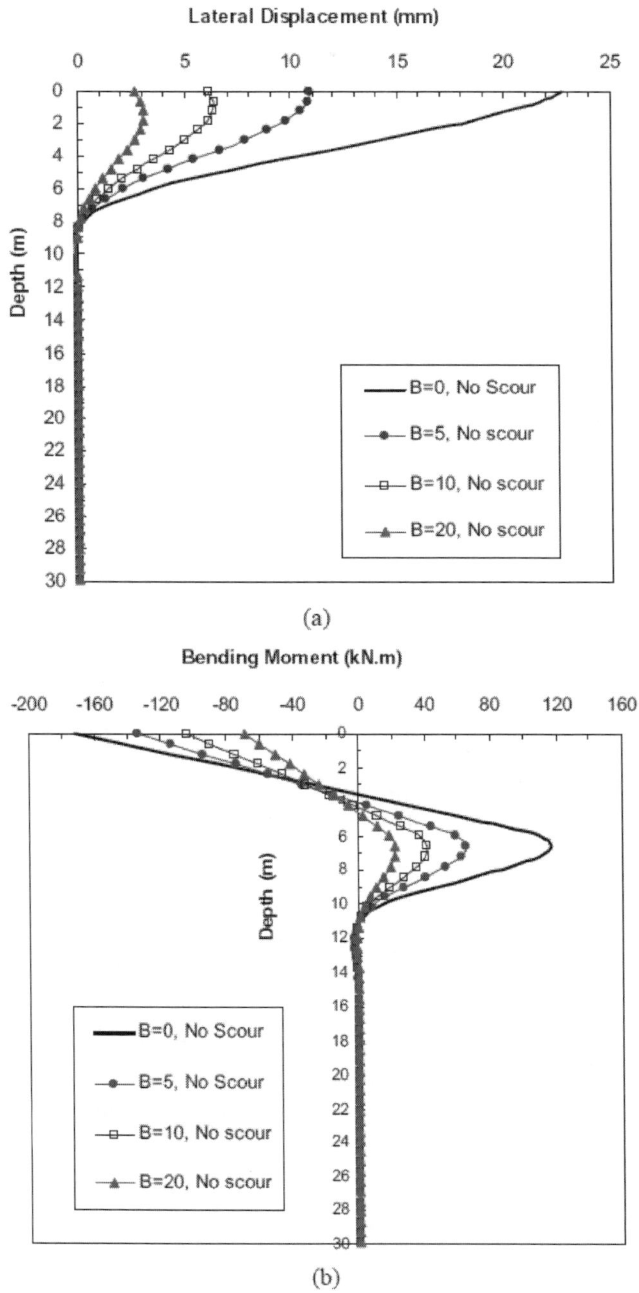

Figure 6. Effect of batter angle on the pile response (F = 100 kN) (a) Shear force along pile length; (b) Bending moment along pile length.

mentioning that bending moment and deflection of positive batter piles are almost similar or slightly higher than that for negative batter piles. This may be attributed to the pile head conditions which were assumed to be fixed in all analyses. Due to scale limitations, only negative piles are presented.

Figure 7 shows that lateral pile group capacity significantly increases by the increase in batter angle. A batter angle of 5° increases the lateral capacity by about 40% over vertical piles (b = 0). The increase in the group capacity is almost linear. The rate of increase in capacity if b ranges between 10° and 20° is higher than that rate if b ranges between 0° and 10°. If scour depth equivalent to twice pile diameter ($d_s/d = 2$) is considered, the group lateral capacity decreases by about 5% to 10%. The rate of change decreases with the increase in batter angle. In another meaning, the scour becomes less significant with the increase in batter angle.

Wave loading is cyclic in nature. In offshore conditions the number of cycles of wave loading may be up to 600 [14]. The cyclic loading causes a gap between the pile and soil interface. Effect of cyclic loading on the behavior of pile groups was examined in this paper. No major difference was found in the results for the cyclic loading conditions compared to static loading conditions. This is because the analysis was performed in sandy soils as sand tends to fill the gap between the pile and soil interface during cyclic loading. For cohesive soils, cyclic loading is expected to have larger impact on the piles compared to static loading.

Caution should be exercised for seismic loading conditions as it has been reported in some publications that batter piles have poor performance during recent earthquakes [16].

4. EFFECT OF SCOUR ON VERTICAL PILE GROUPS

Effect of global scour on vertical pile groups was investtigated. Several parameters were considered including the variation in scour depth. The effect of scour depth combined with other parameters such as different pile arrangement, spacing between piles and pile slenderness ratio were investigated.

The same soil and pile characteristics described in Section 3 were assumed except that all piles are considered to be vertical. All pile groups comprised two piles. No free length was assumed in the analysis of vertical piles.

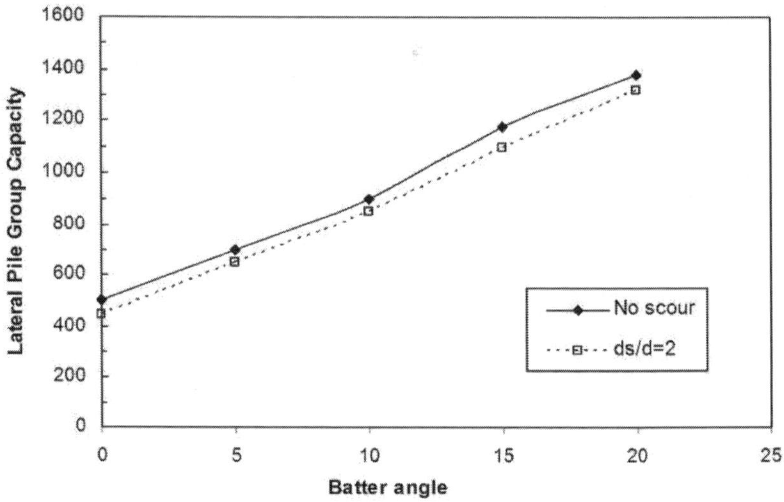

Figure 7. Relationship between pile batter angle and the lateral pile group capacity (d = 0.5 m, L = 30 m, L_f = 5 m, S/d = 3).

Assuming the piles are Grade 4 steel with a yield stress of 450 MPa, the equivalent moment causing the pile to yield is about 1040 kN/m. The ultimate lateral pile capacity corresponds to the load that causes the yield moment.

4.1. Scour Depth around Pile Groups

The scour depth around pile groups is based on the root cause of scour. Piles supporting coastal and marine structures are subjected to scour due to waves, currents and ship propeller jets. A summary of the expected scour depth due to these causes is provided in the following subsections.

4.1.1. Scour Due to Waves and Current

Some experimental research has been performed to examine the scour around pile groups due to waves and currents. Sumer and Fredsøe [2] examined the wave scour around a group of closed spaced piles. Sumer et al. [17] investigated the global and local scour at pile groups exposed to steady currents. Mostafa and Agamy [3] examined the combined effect of waves and current on a group of closed and widely spaced piles. Scour depth to pile diameter ratio (d_s/d) due to waves only can be as high as 1 for Keulegan-Carpenter number (KC) of 13. For pile groups exposed to currents only, total scour (global scour and local scour) increases with the increase in number of piles. The scour depth to pile diameter ratio (d_s/d) can be as high as 2.6 [17].

4.1.2 Scour Due to Ship Propeller Jets

During berthing or de-berthing operations, a ship is in proximity to harbor structures and erosion may occur around piles supporting jetties or relieving platforms and sloping riprap [18]. **Figure 8**shows a sketch of scour around piles supporting jetties or relieving platforms due to ship propeller jets. It is noticed that scour at the front (leading) pile is larger than that at the trailing (aft) pile. If pile spacing increases, scour at the trailing pile decreases as a result of the deposition of scoured sediment taking place at the vicinity of the front pile as can be seen in Figure 8.

Yuksel et al. [18] concluded that maximum scour depth and sand deposition for pile groups vary significantly with densimetric Froude number Fr_d, pile spacing to diameter ratio (S/d) and pile diameter to jet diameter ratio (d/d_o). Scour at the front (leading) pile is larger than that at the trailing (aft) pile. Empirical equations for scour calculations due to ship propeller jets can be found in the literature. Experimental results of Yuksel et al. [18] indicated that maximum scour depth to pile diameter ratio (d_s/d) is about 2.25.

In this paper, normalized scour depth (d_s/d) between 0 and 3 was considered in the analysis to cover the different root causes of scour around piles supporting coastal structures.

4.2. Effect of Pile Arrangement

The analysis considered two main pile groups. The pile groups were assumed to be arranged side-by-side (i.e., q = 90°) or tandem arrangement (i.e., q = 0°) where q is the angle between the lateral load and the line connecting the two piles and S is the centre line to centre line spacing between piles. **Figure 9** shows a general sketch indicating a pile location within a pile group subjected to lateral loading. The piles were assumed to be spaced three times the pile diameter (i.e., S/d = 3). Pile-soil-pile interaction was considered in the analysis. The reduction factors generated in the program Group [9] were used in this study. The length to diameter ratio (L/d) was assumed to be 60 so that pile embedment length has no impact on the results.

Figure 10 shows the effect of normalized scour depth (d_s/d) on the lateral capacity of a group of two piles with side-by-side arrangement and tandem arrangement. Figure 10(b) indicates that the reduction in pile group capacity of groups with side-by-side arrangement is higher than that for groups with tandem arrangement. A relatively small scour depth ($d_s/d = 1$) leads to a reduction in lateral group capacity ranging between about 12.5% and 15% for piles with tandem arrangement and side-by-side arrangement, respectively. A scour depth of three pile diameter (i.e., $d_s/d = 3$) causes a reduction in the lateral group capacity ranging between 29% and 35% for piles with tandem arrangement and side-by-side arrangement, respectively. The rate of reduction in lateral group capacity if d_s/d varies between 0 and 1 is slightly higher than that if d_s/d is between 1 and 3.

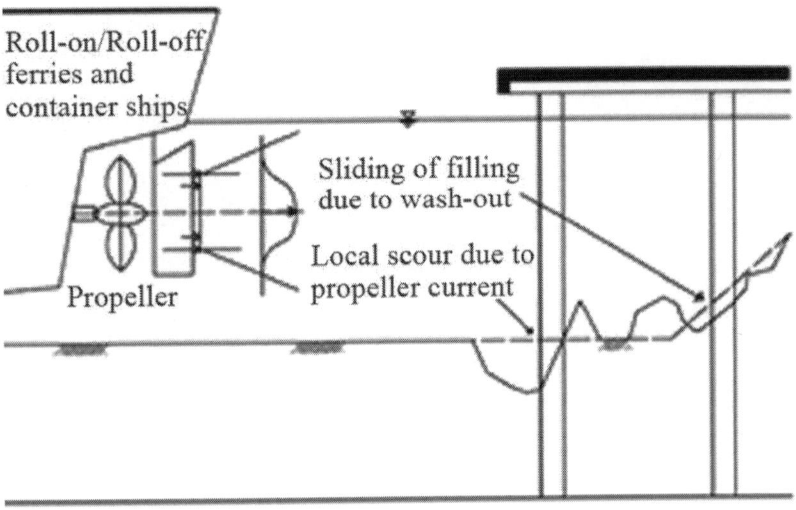

Figure 8. Erosion of seabed due to ship propeller jets (Chin et al., 1996 [19]).

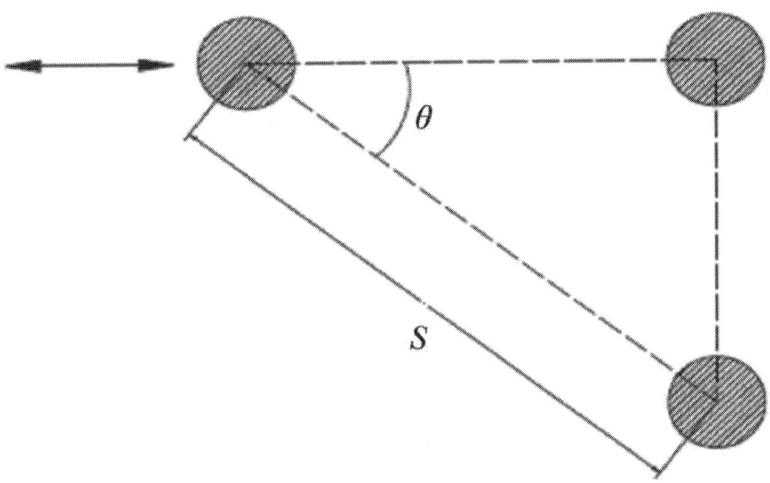

Figure 9. The location of a pile in pile group under lateral loading.

Figure 11 shows load-displacement curves for groups of two piles with tandem arrangement (q = 0°) and sideby-side arrangement (q = 90°) neglecting and considering scour. Load-displacement curves for single pile computed using program LPILE were plotted for comparison. It can be seen that the load-displacement curve for single

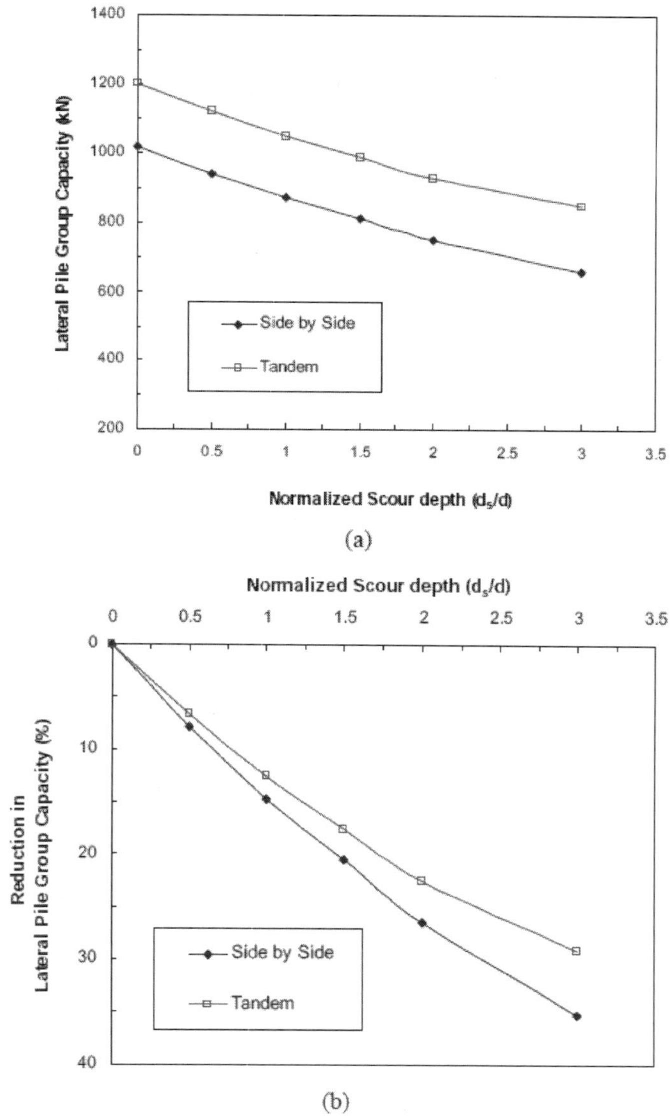

Figure 10. Relationship between normalized scour depth and lateral pile capacity (S/d = 3, L = 30 m, L/d = 60).

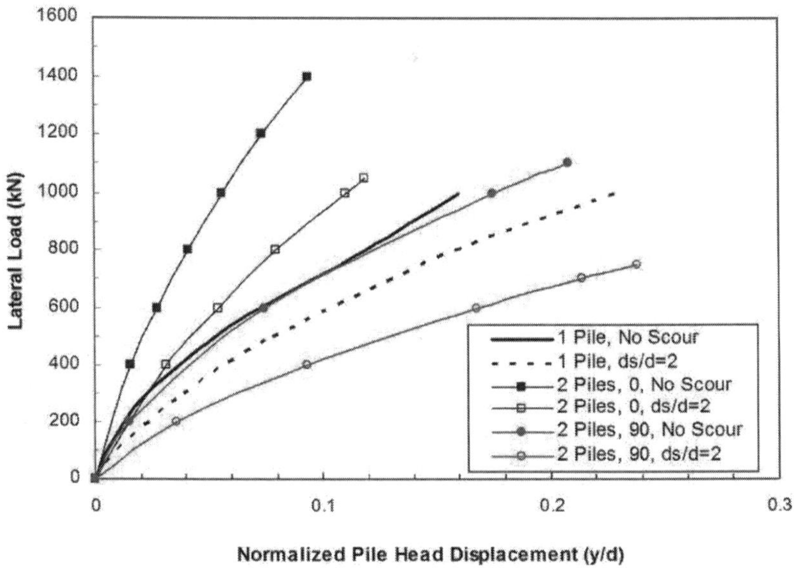

Figure 11. Load-displacement curves for single pile and pile groups with different arrangement neglecting and considering scour (S/d = 3, L = 30 m, L/d = 60).

pile is very similar to that for two piles with side-by-side arrangement. This is because the side-by-side reduction factor due to group interaction is negligible if the spacing is greater than three times the pile diameter [Group] [9]. For the same lateral loading, pile group with tandem arrangement has less displacement than that for pile group with side-by-side arrangement. For relatively small lateral loads on the group (F < 100 kN) no major differences in the normalized pile head displacement (y/d) are found due to pile group arrangement. The impact of pile arrangement is more pronounced with the increase in lateral loading due to the nonlinearity in the pile-soil system. For single pile and pile groups with different arrangements, scour increases the pile head displacement. For the same lateral loading, piles with tandem arrangement even when scour is considered has lower displacement than the piles with side-by-side arrangement with no scour. This is due to the greater reduction in group efficiency for side-by-side arrangement compared to the reduction in group efficiency for tandem arrangement.

Figure 12 shows lateral displacement and bending moment along pile length when a group of two piles is subjected to lateral load (F) of 200 kN. For the sake of comparison, single pile subjected to lateral load of 100 kN was computed and plotted. It is noted that displacement of piles with side-by-side arrangement (q = 90°) is higher than that for single pile and pile

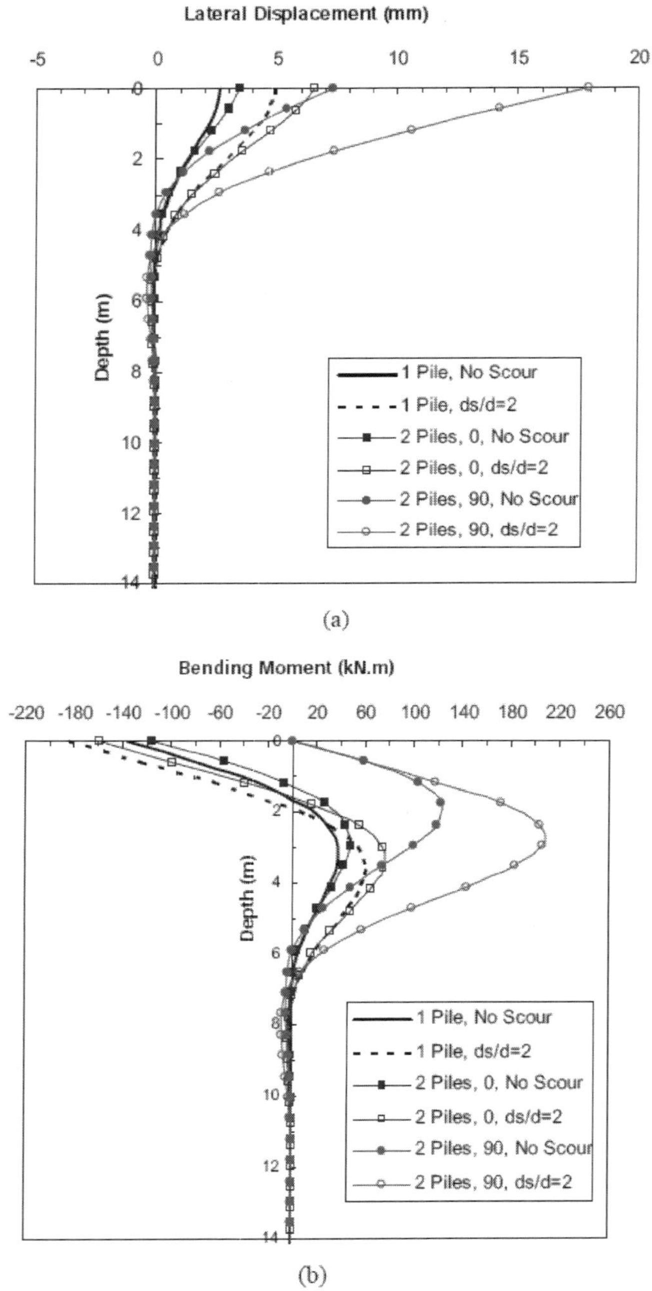

Figure 12. Effect of pile arrangement on lateral response along pile length, (a) Lateral displacement; (b) Bending moment (F = 200 kN, S/d = 3, L = 30 m, L/d = 60).

group with tandem arrangement (q = 0°). Scour depth of twice pile diameter increases the head displacement of pile group with side-by-side arrangement from 7.4 mm to 18 mm (i.e., increase by about 114%). For piles with tandem arrangement, the displacement at pile head increases due to scour from about 3.4 mm to 6.5 mm (i.e., increase by about 91%). For the single pile, scour increased the pile head displacement from about 2.6 mm to about 4.9 mm (i.e., increase by about 88%). Under the same loading conditions, the displacement of pile group is higher than that for single pile. This is attributed to the pile-soil-pile interaction. The scour effect on pile group with side-byside arrangement is more significant than that for single piles and pile group with tandem arrangement. The bending moment for piles with side-by-side arrangement is larger than that for single pile and pile group with tandem arrangement.

4.3. Effect of Pile Spacing

The effect of centre line to centre line pile spacing considering pile-soil-pile interaction was evaluated. Spacing to diameter ratio (S/d) of 1, 3 and 5 for pile groups with tandem and side-by-side arrangements was considered. The impact of scour is also investigated.

When piles are closely spaced, the shear failure planes resulting from the movement of each pile will overlap and the ultimate resistance for a pile in a group may be less than that of a single pile. This is called "shadow effect" or "pile-soil-pile interaction" which influences the efficiency of individual piles in a group. Reduction factors generated in the program Group [9] were used in this study. These reduction factors are based on many laboratory and field experiments collected from the literature.

Brown et al. [20] introduced the concept of the pmultiplier. This concept represents the response of the group to lateral loading in terms of the response of an assembly of single piles with the soil reaction modeled using p-y curves adjusted using a "p-multiplier" where p is the lateral soil reaction and y is lateral pile displacement. The p-multiplier assumes a different value that depends on whether a pile is in a leading or in a trailing position and the angle between the line connecting two piles and the load direction.

Figure 13 shows the relationship between normalized scour depth and lateral pile group capacity taking into consideration the reduction in group efficiency due to pile-soil-pile interaction. It is evident that the lateral pile group capacity for the two piles with tandem arrangement is higher than that of the two piles with side-by-side arrangement by about 18% to 32%. For closely spaced piles (i.e., S/d = 1) the group capacity is reduced by about 11% and 5% for case of side-by-side arrangement and tandem arrangement respectively. A scour depth equivalent to three times pile diameter decreases the lateral group capacity by about 32% to 35% for groups with side-by side arrangement. The same scour depth reduced the capacity by about 27% to 30% for groups with tandem arrangement.

It should be noted that for tandem arrangement, the ultimate lateral capacity of trailing pile is lower than that of leading pile assuming the scour depth is the same at leading and trailing piles. Therefore, lateral load that causes yield moment of trailing pile was considered in **Figure 13**.

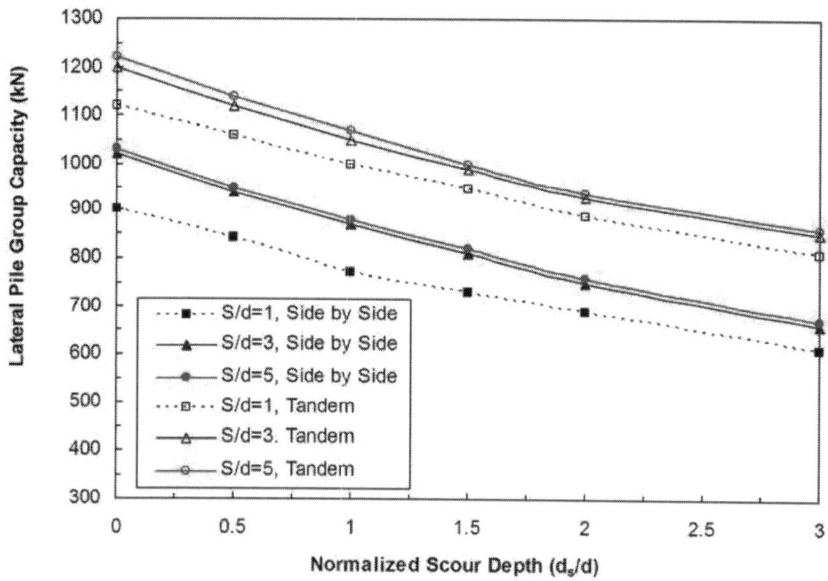

Figure 13. Normalized scour depth versus lateral pile group capacity for different pile spacing and arrangement.

Figure 14 shows load-displacement curves for two pile group with side-by-side arrangement considering different pile spacings. Impact of scour was also considered by assuming a scour depth equivalent to twice the pile diameter. **Figure 14** indicates that pile displacement increases significantly when the piles are closely spaced and scour is considered. As expected, the ultimate group capacity increases with the increase in pile spacing and if scour is neglected. The effect of scour is more pronounced with the increase in lateral loading. As an example, a lateral load of 600 kN on a group with S/d = 1 causes a displacement (y/d) of 0.11 and 0.23 if scour is neglected and considered, respectively. For the same lateral load applied on a group of widely spaced piles (S/d = 5), y/d is computed to be 0.074 and 0.17 if scour is neglected and considered, respectively. For the same pile spacing ratio, the displacement increases by about 100% to 130% due to scour. If scour is neglected and pile spacing ratio is varied, the displacement increases by about 7% if S/d is reduced from 5 to 3 and the displacement increases by about 40% if S/d is reduced from 3 to 1. This is due to the pile-soil-pile interaction. If scour is neglected, the ultimate lateral group capacity is reduced from 1050 kN if S/d = 5 to 900 kN if S/d = 1. This corresponds to a reduction of about 14% due to pile-soil-pile interaction. For the same S/d ratio, the ultimate lateral capacity reduces by about 23% to 29% due to the scour effect.

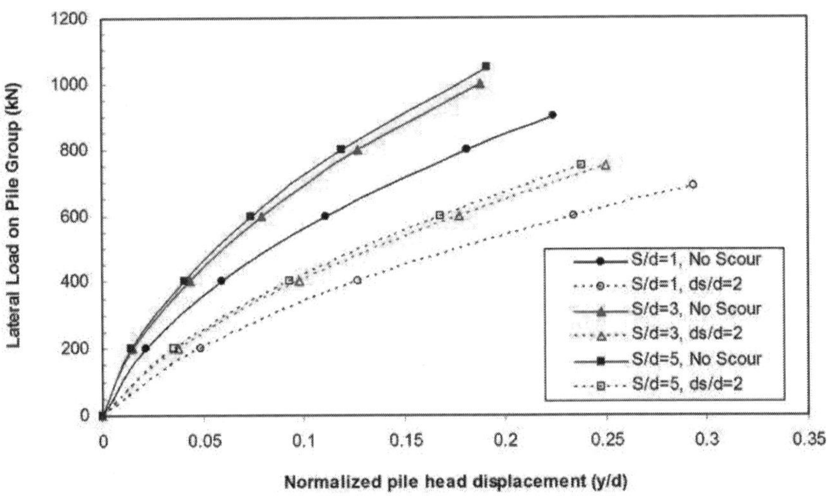

Figure 14. Load-displacement curves for pile group with different pile spacing with and without scour (side-by-side arrangement, L/d = 60).

Therefore, it is concluded that the effect of scour on lateral group capacity and displacement is more pronounced than the effect of pile-soil-pile interaction.

4.4. Effect of Pile Slenderness Ratio

Pile length to diameter ratio (slenderness ratio) may be critical for piles supporting coastal structures under scoured conditions. To examine the effect of pile slenderness ratio, L/d ratios of 5, 7.5, 10, 12.5, 15, 20 and 40 were considered. Pile slenderness ratio for some dolphins or suction piles may be as low as 5. The pile spacing ratio (S/d) was kept constant as 3.

Groups of two piles with side-by-side arrangement were considered as this is a more critical case than groups with tandem arrangement.

Figure 15 shows the relationship between normalized scour depth and lateral pile capacity for different pile slenderness ratio. It is noted that for L/d = 5, the rate of decrease in lateral pile group capacity between $d_s/d = 0$ to 0.5 is higher than that rate if d_s/d varies between 0.5 and 3. For L/d = 7.5 and 10, the rate of decrease in lateral group capacity between $d_s/d = 0$ and 1 is higher than that rate if d_s/d varies between 1 to 3.

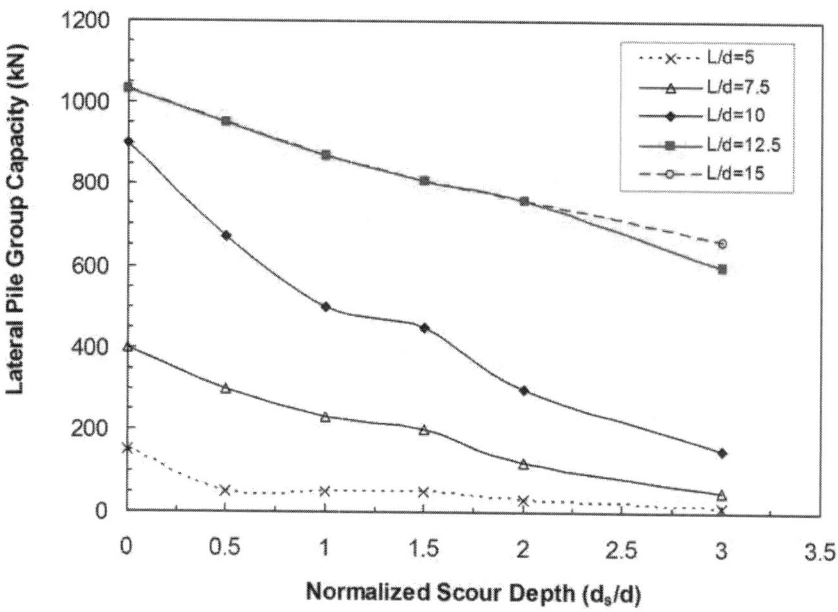

Figure 15. Lateral pile group capacity versus normalized scour depth (ds/d) considering different pile slenderness ratio (L/d) (S/d = 3, side-by-side arrangement).

It is also noted from **Figure 15** that percentage decrease in lateral pile group capacity (PDC) remains almost constant after the pile slenderness ratio is greater than about 12.5. In another meaning, for pile groups exposed to scour, increasing L/d ratio over 12.5 will not add much difference to the group capacity. Accordingly, the decrease in group capacity due to scouring will be more serious for short piles. This is somewhat similar to a conclusion drawn by Ni et al. [8] who concluded that for single piles, PDC values remain almost constant after L/d is greater than 10.

5. SCOUR PROTECTION

As discussed in this paper, pile groups subjected to scour due to waves, current and ship propeller jets may be exposed to a scour depth up to 2.6 pile diameter. A value of scour depth to diameter ratio (d_s/d) of 2 is often used in practice. It is important for designers and practitioners to know when scour protection around piles supporting coastal structures may be necessary and when it is not.

According to Mostafa and Agamy [3] and Abdeldayem et al. [21], pile groups with side-by-side arrangement causes more scour than groups with tandem arrangement. Scour depth for some cases of pile groups with side-by-side arrangement increases as much as about two times more than its magnitude for the case of single pile according to Mostafa and Agamy [3] and Sumer et al. [17]. Accordingly, for a jetty or berthing structure supported on piles exposed to scour due to ship propeller jets, the leading pile row, which is arranged side-by-side, is expected to suffer more from scour.

From a geotechnical prospective, the ultimate lateral capacity for piles arranged side-by-side is lower than that for piles with tandem arrangement. The side-by-side arrangement is exposed to higher lateral deflections and bending moments. Therefore, the combined effect of scour and pile-soil-pile interaction for closely spaced piles makes the side-by-side arrangement the worst case scenario in terms of significant reduction in pile capacity. This scenario may be even worse if the loading arm or free length (distance between jetty deck level and seabed) is high. In this case, scour protection and frequent monitoring of erosion around the piles may be required.

For raked piles, as scour is less significant when the batter pile angle increases, scour protection is not necessary depending on the required lateral pile capacity.

For short piles, consideration should be given to using scour protection measures such as riprap or geotextile. Alternatively, consideration should be given to increasing the pile length so that the slenderness ratio is about 12.5. For already constructed piles with slenderness ratio less than 12.5, consideration should be given to using a suitable scour protection measure.

It should be noted that this section provides general recommendations on when to use scour protection. The decision to use scour protection around piles should be taken based on the pile, soil and loading conditions.

6. CONCLUSIONS

Some design aspects of pile groups supporting marine structures founded on cohesionless soils were studied using numerical modeling. These aspects are scour depth, lateral static loading, batter pile angle, pile group arrangement, pile spacing and pile slenderness ratio. For the input data assumed in the analyses, the following conclusions can be drawn:

1) A small increase in batter pile angle increases the lateral pile capacity significantly. A batter angle of 5° increases the lateral capacity by about 40% over vertical piles.

2) Scour depth becomes less significant with the increase in pile batter angle.

3) Pile groups with side-by-side arrangement experience higher displacement and bending moments compared to single piles and pile groups with tandem arrangement due to the combined effect of scour depth and pilesoil-pile interaction.

4) The ultimate lateral capacity for pile groups with side-by-side arrangement is lower than that for pile groups with tandem arrangement. For spacing to diameter ratio of 3, a scour depth of three pile diameter (i.e., $d_s/d = 3$) causes a reduction in the lateral group capacity ranging between 29% and 35% for piles with tandem arrangement and side-by-side arrangement, respectively.

5) The rate of reduction in lateral group capacity if d_s/d varies between 0 and 1 is slightly higher than that if d_s/d is between 1 and 3.

6) For relatively small lateral loads on the group ($F < 100$ kN), no major differences in the normalized pile head displacement (y/d) are found due to the pile group arrangement. The impact of pile arrangement is more pronounced with the increase in lateral loading due to nonlinearity in the pile-soil system.

7) For the same lateral loading, piles with tandem arrangement even when scour is considered has lower displacement than piles with side-by-side arrangement with no scour.

8) The bending moment for piles with side-by-side arrangement is larger than that for single pile and pile group with tandem arrangement.

9) The effect of scour on lateral group capacity and displacement is more pronounced than the effect of pilesoil-pile interaction.

10) The percentage decrease in lateral pile group capacity (PDC) remains almost constant after the pile slenderness ratio is greater than about 12.5.

11) More attention and consideration should be given to scour protection around piles especially if the piles are closely spaced, arranged side-by-side and if slenderness ratio is less than 12.5.

REFERENCES

1. B. M. Sumer, J. Fredsøe and N. Christiansen, "Scour around a Vertical Pile in Waves," Journal of Waterway, Port, Coastal and Ocean Engineering, ASCE, Vol. 118, No. 1, 1992, pp. 15-31.
2. B. M. Sumer and J. Fredsøe, "Wave Scour around Group of Vertical Piles," Journal of Waterway, Port, Coastal and Ocean Engineering, ASCE, Vol. 124, No. 5, 1998, pp. 248-255.
3. Y. E. Mostafa and A. F. Agamy, "Scour around Single Pile and Pile Groups Subjected to Waves and Currents," International Journal of Engineering Science and Technology, IJEST, Vol. 3, No. 11, 2011, pp. 8160-8178.
4. C. Lin, C. Bennett, J. Han and R. L. Parsons, "Integrated Analysis of the Performance of Pile-Supported Bridges under Scoured Conditions," Engineering Structures, Vol. 36, 2012, pp. 27-38.
5. Y. N. Kishore, S. N. Rao and J. S. Mani, "The Behaviour of Laterally Loaded Piles Subjected to Scour in Marine Environment," KSCE Journal of Civil Engineering, Vol. 13, No. 6, 2009, pp. 403-406.
6. C. Lin, C. Bennett, J. Han and R. L. Parsons, "Scour Effects on the Response of Laterally Loaded Piles Considering Stress History of Sand," Computers and Geotechnics, Vol. 37, No. 7-8, 2010, pp. 1008-1014.
7. Y. E. Mostafa, "Effect of Local and Global Scour on Lateral Response of Single Piles in Different Soil Conditions," Engineering, Vol. 4, No. 6, 2012, pp. 297-306.
8. S. Ni, Y. Huang and L. Lo, "Numerical Investigation of the Scouring Effect on the Lateral Response of Piles in Sand," Journal of Performance of Constructed Facilities, Vol. 263, 2012, pp. 320-325.
9. Ensoft Inc., "GROUP, A Program for Analyzing a Group of Piles Subjected to Axial and Lateral Loading," Version 7.0, Technical Manual, Austin, 2006.
10. R. Whitehouse, "Scour at Marine Structures, A Manual for Practical Application," Thomas Telford Publications, London, 1998.

11. M. F. Randolph, "PIGLET. A Program for Analysis and Design of Pile Groups," Version 5.1, Technical Manual, The University of Western Australia, Perth, 2004.

12. L. M. Zhang, M. C. McVay, S. J. Han, P. W. Lai and R. Gardner, "Effects of Dead Loads on the Lateral Response of Battered Pile Groups," Canadian Geotechnical Journal, Vol. 39, No. 3, 2002, pp. 561-575.

13. T. Tazoh1, M. Sato, J. Jang and G. Gazetas, "Centrifuge Tests on Pile Foundation-Structure Systems Affected by Liquefaction-Induced Soil Flow after Quay Wall Failure," The 14th World Conference on Earthquake Engineering, Beijing, 12-17 October 2008.

14. S. S. Rajashree and T. G. Sitharam, "Nonlinear FiniteElement Modeling of Batter Piles under Lateral Load," Journal of Geotechnical and Geoenvironmental Engineering, Vol. 127, No. 7, 2001, pp. 604-612.

15. G. P. Tschebotarioof, "The Resistance to Lateral Loading of Single Piles and Pile Groups," Special Publication No. 154, ASTM, 1953, pp. 38-48

16. R. E. Harn, "Displacement Design of Marine Structures on Batter Piles," 13th World Conference on Earthquake Engineering, Vancouver, 1-6 Augsut 2004, Paper No. 543.

17. B. M. Sumer, K. Bundgaard and J. Fredsøe, "Global and Local Scour at Pile Groups," Proceedings of the Fifteenth International Offshore and Polar Engineering Conference, Seoul, 19-24 June 2005, pp. 577-583.

18. A. Yuksel, Y. Celikoglu, E. Cevik and Y. Yuksel, "Jet Scour around Vertical Piles and Pile Groups," Ocean Engineering, Vol. 32, No. 3-4, 2005, pp. 349-362.

19. C. Chin, Y. Chiew, S. Y. Lim and F. H. Lim, "Jet Scour around Vertical Pile," Journal of Waterway, Port, Coastal and Ocean Engineering, Vol. 122, No. 2, 1996, pp. 59-67.

20. D. A. Brown, C. Morrison and L. C. Reese, "Lateral Load Behavior of Pile Group in Sand," Journal of Geotechnical Engineering, ASCE, Vol. 114, No. 11, 1988, pp. 1326- 1343.

21. A. Abdeldayem, G. Elsaeed and A. Ghareeb, "Effect of Pile Group Arrangements on Local Scour Using Numerical Models," Advances in Natural and Applied Sciences, Vol. 52, 2011, pp. 141-146.

10
CHAPTER

Detection of Nonlinearity Effects in Structural Integrity Monitoring Methods for Offshore Jacket-Type Structures Based on Principal Component Analysis

A. Mojtahedi [a,*], M.A. Lotfollahi Yaghin [a], M.M. Ettefagh [b], Y. Hassanzadeh [a], M. Fujikubo [c]

[a]*Department of Civil Engineering, University of Tabriz, Tabriz, Iran*

[b]*Department of Mechanical Engineering, University of Tabriz, Tabriz, Iran*

[c]*Department of Naval Architecture and Ocean Engineering, Graduate School of Engineering, Osaka University, Osaka, Japan*

ABSTRACT

The detection of changes in the dynamic behavior of structures is an important issue in structural safety assessment. The development of detection methods assumes greater significance in the case of offshore platforms because the inherent problems are compounded by the harsh environment. Here, we describe an instrumented physical model for the structural health monitoring of an offshore jacket-type structure and the results of tests in several different damage scenarios. In a comparative investigation of two different methods, we discuss the difficulties of implementing damage detection techniques for complex structures, such as offshore platforms. The combined algorithm of a fuzzy logic system and a model updating method are briefly discussed, and a method based on stochastic autoregressive moving average with exogenous input is adopted for the structure. The consideration of uncertainties and the effects of nonlinearity were major objectives. So, the methods were also investigated based on the test scenarios consisting of the physical model with a geometric nonlinearity. The principal component analysis method was utilized for the detection of nonlinearity in the recorded data. The results show that the developed methods are suitable for damage classification, but the quality of the acquired signals must be considered an important factor influencing successful classification. The development of these methods may be extremely useful, as such technologies could be applied for offshore platforms in service, enabling damage detection with fewer false alarms.

Keywords: Offshore structure; Damage identification; Parametric odel; Experimental verification

1. INTRODUCTION

Structural health monitoring (SHM) and fault-detection processes are essential for warranting structural safety performance during the service life of a structure. Research on vibration-based damage identification using changes in output signals from the structure has expanded rapidly in the last decades. Doubling et al. [1] published a detailed, state-of-the-art review on vibration-based damage identification methods. During the 1970s and 1980s, the offshore structures industry began to study damage identification problems in offshore structures, particularly in jacket platforms [2]. Located at sea, these structures are in direct contact with the environment, and any unexpected failure can be calamitous in nature. The recent disastrous oil spill in the Gulf of Mexico emphasizes the importance of damage identification problems for the offshore oil drilling and marine industries. Concern over structural failure feedback effects on environmental pollution became more pronounced in Japan during the recent nuclear incident involving the emission of radiation into seawater around the plant.

In recent years, several researchers have studied damage detection in offshore jacket platforms. Nichols [3] discussed structural monitoring in offshore structures and described a fault-detection algorithm using phase space methods. This work provides a detailed overview of the background research and serves as a good reference in this field.

Elshafey et al. [4] and [5] examined damage diagnosis in offshore jacket platforms using mode shapes and a combined method of random decrement signature and neural networks. Li et al. [6] proposed the 'modal strain energy decomposition method' to detect damage in offshore structures using changes in mode shapes based on modal testing. Liu et al. [7] investigated a method based on the Hilbert–Huang transform method for modal parameter identification in offshore platform structures.

Roitman et al. [8]presented a methodology using frequency response functions to determine the actual conditions of the structure to detect damage.

As evidenced in the literature, due to the many practical problems encountered in such methods, e.g., structural complexity, load and operational conditions and difficulties caused by various types of uncertainties, efforts at further developing these techniques for offshore platforms were largely abandoned by the early 1980s [9]. For this reason, in subsequent years, few researchers have addressed damage detection in offshore jacket platforms. In addition, despite of the aforementioned effort, there has been no work directly focusing on the effects of uncertainties as the main scope of the study in an effort to circumvent the major problem and applying such techniques to offshore platforms.

In this study, the problems of uncertainties and the effects of nonlinearity were considered to be the main objectives in developing and evaluating a robust damage detection system. The realization of this type of nonlinearity may depend on the output excitation for the structure. When the excitation levels increase, the tall structure shows the large displacements and a nonlinear geometric effect is activated. These concepts are investigated by the adaptation of two methods based on different standpoints: experimental modal analysis and time-capture data processing. In addition, one implicit objective of this study is to discuss the inherent difficulties of implementing SHM techniques for complex structures, such as offshore platforms. The development of such methodologies would be extremely useful in enabling technologies that can be applied to offshore platforms in service to provide improved damage detection with fewer false alarms.

In this work, a physical platform model was constructed for this purpose. To allow for the empirical evaluation of the proposed damage assessment method, the model was designed so that several damage scenarios can be introduced by stiffness reduction in selected structural members. Experimental vibration tests were conducted using data based on operational modal analysis with exogenous forces (OMAX) [10].

First, a combined algorithm incorporating a fuzzy logic (FL) system and a model updating method is briefly described. The method uses a modal parameter in the frequency domain with a consideration of the uncertainties associated with ambiguous damage states. This technique is demonstrated to be effective for diagnosing degradation and quantifying damage. However, investigation of the experimental results revealed an unsatisfactory classification of some damage scenarios [11].

Therefore, a method using extracted features from an output time-series signal was adopted for the structure. This approach is based on a stochastic autoregressive moving average with exogenous input (ARMAX) method.

One advantage of this method is the direct handling of the time-series data, whereas the previous frequency domain method required the measured time data to be transformed from the time domain to the frequency domain before feature extraction processing. Moreover, the first method is based on a model updating technique that requires the construction of a numerical model. Although the first method offers the advantage of identifying the damage location and possibly calculating the remaining life time, it generally requires more computational time [12]. The second method was validated using the same experimental data; however, despite the noted advantages, similar to the first method, it failed to correctly diagnose several cases.

The concepts of these two methods were developed based on the linear geometric characteristics of related system responses; finally, the nonlinearity in the dynamic system was evaluated. Here, principal component analysis (PCA) and the concept of subspace angle were adopted for the detection of nonlinearity in the recorded data. After a reasonable investigation, the observed error rates of the methods were attributed to neglecting some nonlinear activation. The observed results are discussed in detail in Section 5.

2. PROBABILISTIC DAMAGE IDENTIFICATION APPROACHES IN THIS STUDY

2.1. Method based on experimental modal features: Combined model updating and FL system

This method consists of a combined strategy of experimental and numerical modal analysis. An FL system is used as the basis to demonstrate the state of the system considering the uncertainties associated with ambiguity and imprecision. Database construction for the related FL system requires many structural response data series that must be collected during numerous fault scenarios. It is impossible to define each of these cases in an empirical manner; therefore, a numerical finite element (FE) model must be used. Of course, the numerical simulation is not itself a determinate process in the development of SHM methodologies, and therefore, final reliable judgments should be based on experiments [13]. However, differences are always found between the numerically and experimentally identified dynamic characteristics due to various types of modeling errors, which can affect the accuracy of the damage detection process. Hence, updating the initial FE model and the adjustment of several model parameters are required to minimize the model error based on the experimental results [14]. Although both experimental and numerical methods contain errors in the measured frequencies, during the damage detection process based on the FE model updating, the experimental results are more acceptable and are thus used as the objective. Therefore, the FE model is updated with the aim of aligning the results of the numerical natural frequencies to the experimental values. A brief description of the method follows.

2.1.1. Experimental modal analysis

Experimental modal analysis is identified simply as a process for describing a structure in terms of its dynamic properties, such as frequencies, damping and mode shapes. Modal analysis is basically the study of the natural characteristics of a structure. The methods can be classified into different groups, such as operational modal analysis (OMA), experimental modal analysis (EMA) and OMAX [10] and [15].

In the mechanical and signal processing laboratories, the measured responses can be obtained from the shaker or hammer impact tests (Fig. 1). From a theoretical standpoint, there is no difference between these methods, but there are many different practical considerations when performing the shaker or impact tests [16]. The modal analysis techniques used for damage detection and the structural monitoring process in this study are described in Section 4.

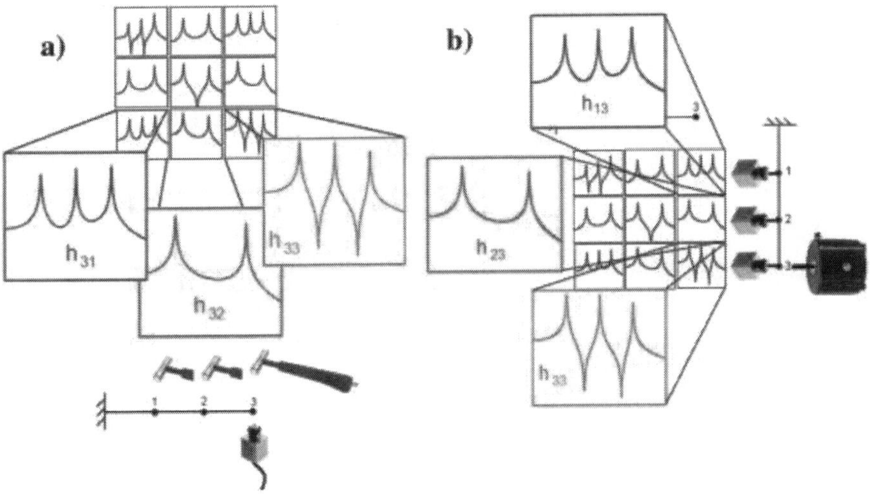

Figure. 1. Experimental modal analysis: (a) moving impact test, (b) moving response test [16].

2.1.2. Numerical modal analysis

FE software packages can be used to analyze the initial FE model according to the specifications of the structure. Here, the term "initial" implies that the FE model is considered to be the basis for model updating. The material properties and the boundary conditions are very important aspects of the modeling process. The equation of motion for a multi degree damped structural system under support excitation can be expressed as follows:

$$\mathbf{M}\ddot{\mathbf{X}}(t) + \mathbf{C}\dot{\mathbf{X}}(t) + \mathbf{K}\mathbf{X}(t) = \mathbf{F}(t) \tag{1}$$

where **M**, **C** and **K** are the mass, damping and stiffness matrixes and $\mathbf{X}(t)$, $\dot{\mathbf{X}}(t)$ and $\ddot{\mathbf{X}}(t)$ denote the nodal displacement, velocity and acceleration vectors of the structure, respectively. In most damage detection problems, the method is not affected by damping, so the undamped natural frequencies are considered to be the desired extracted features. Thus, the following standard Eigenvalue problem must be solved:

$$\det|[\mathbf{K}]-[\mathbf{M}]\omega^2|=0 \tag{2}$$

2.1.3. FE model updating based on the modal assurance criterion (MAC)

The updating of the initial FE model is necessary to minimize the numerical model error according to the experimental signatures. The concepts of the "Modal Assurance Criterion (MAC)" method can be applied for this purpose. Concisely, it can be explained as follows.

Eq. (3) defines a vector of parameters related to modal properties:

$$w = \begin{Bmatrix} w_1 \\ \cdots \\ w_2 \end{Bmatrix} \tag{3}$$

The parameters in the above equations are defined below:

$$w_1 = \left\{ \begin{array}{c} \vdots \\ \left(\frac{\omega_a}{\omega_e}\right)_i \\ \vdots \end{array} \right\}, \; w_2 = \left\{ \begin{array}{c} \vdots \\ \mathrm{MAC}_i \\ \vdots \end{array} \right\} \; \text{and} \; \mathrm{MAC}_i = \left[\frac{\left(\varphi_e^T \varphi_a\right)^2}{\left(\varphi_e^T \varphi_e\right)\left(\varphi_i^T \varphi_a\right)} \right]_i \tag{4}$$

where ω_i and φ_i are the *i*th Eigenvalue and mode shape, respectively, and the subscripts a and e denote the analytical and corresponding experimental values. Using the first-order Taylor series, we obtain the following:

$$w_e = w_a + T\Delta x + \varepsilon \tag{5}$$

where w_e and w_a are the experimental and analytical function vectors, T is the design sensitivity matrix of w_a, $x\Delta x$ are the changes in x for the least squares minimization, and ε is a residual vector. The least squares solution for Δx to minimize $\varepsilon T \varepsilon$ is:

$$\Delta x = \left(T^T T\right)^{-1} T^T \Delta w = \left(T^T T\right)^{-1} T^T (w_e - w_a) = \left(T^T T\right)^{-1} T^T (\{1\} - w_a) \tag{6}$$

where the design sensitivity matrix modal functions of the Eigenvalue and the Eigenvector can define as follow:

$$T = \begin{bmatrix} \frac{\partial w_{1a}}{\partial x} \\ \cdots \\ \frac{\partial w_{2a}}{\partial x} \end{bmatrix} = \begin{bmatrix} \frac{\partial[\omega_a/\omega_e]}{\partial x} \\ \cdots \\ \frac{\partial[\mathrm{MAC}]}{\partial x} \end{bmatrix} \tag{7}$$

Eq. (4) is rearranged as follows:

$$\mathrm{MAC}_i = \left[\frac{\left(\varphi_e^T \varphi_a\right)^2}{\left(\varphi_e^T \varphi_e\right)\left(\varphi_a^T \varphi_a\right)} \right]_i \equiv \left(\frac{\tau}{\beta}\right)_i \tag{8}$$

where $\tau_i \equiv (\varphi_e^T \varphi_a)_i^2$ and $\beta_i \equiv (\varphi_e^T \varphi_e)_i (\varphi_a^T \varphi_a)_i$. If the MAC with value of 1 indicates perfect correlation, then the partial derivative of MAC$_i$ with respect to the design variable *xj* can be written as follows:

$$\frac{\partial \mathrm{MAC}_i}{\partial x_j} = \frac{\beta_i \frac{\partial \tau_i}{\partial x_j} - \tau_i \frac{\partial \beta_i}{\partial x_j}}{\beta_i^2}$$

$$\frac{\partial \tau_i}{\partial x_j} = \frac{\partial (\varphi_e^T \varphi_a)_i^2}{\partial x_j} = 2(\varphi_e^T \varphi_a)_i \left(\varphi_e^T \frac{\partial \varphi_a}{\partial \varphi_j} \right)_i$$

$$\frac{\partial \beta_i}{\partial x_j} = (\varphi_e^T \varphi_e)_i \left(\frac{\partial \varphi_a^T}{\partial x_j} \varphi_a + \varphi_a^T \frac{\partial \varphi_a}{\partial x_j} \right)_i = 2(\varphi_e^T \varphi_e)_i \left(\varphi_a^T \frac{\partial \varphi_a}{\partial \varphi_j} \right)_i \tag{9}$$

A more detailed explanation of this method and a design sensitivity analysis can be found elsewhere [17] and [18].

2.1.4. FL system and classification of damage scenarios

The concepts of the FL system are used to determine the mapping between the deterioration states, including the severity and location of damages and their modal signatures. In this study, the diagonal bracing members in the vertical plane of the physical model are considered to be the members susceptible to damage, as explained in Section 4. The damage severity is defined by the percent damage parameter *D* [19]:

$$D_k^{(l)} = \left\{ \begin{array}{l} \Delta E_1^{(l)} = 100 \dfrac{E_{\text{undamaged}} - E_{1\,\text{damaged}}^{(l)}}{E_{\text{undamaged}}} \\ \Delta E_2^{(l)} = 100 \dfrac{E_{\text{undamaged}} - E_{2\,\text{damaged}}^{(l)}}{E_{\text{undamaged}}} \\ \vdots \\ \Delta E_p^{(l)} = 100 \dfrac{E_{\text{undamaged}} - E_{p\,\text{damaged}}^{(l)}}{E_{\text{undamaged}}} \end{array} \right\} \tag{10}$$

In this equation, E is the Young's modulus of the material, and the subscripts l and k refer to each structural member and its associated damage intensity, respectively. The damage intensities are classified by the linguistic expressions "slight damage", "moderate damage" and "severe damage" for the ranges of 12.5–37.5%, 12.5–62.5% and 62.5–87.5%, respectively. The parameter p depends on the intervals chosen for increments in ΔE in the range of related damage intensity (in this study, $p = 3$).

The features known as "measurement deltas" (MDs) are the difference between the Eigenvalues of the undamaged and the damaged structure written in a non-dimensional form:

$$z_k^{(l)} = \left\{ \begin{array}{ccc} \cdots & \Delta\omega_{j_1} = 100\dfrac{\omega_{j\ \text{undamaged}} - \omega_{j_1\ \text{damaged}}}{\omega_{j\ \text{undamaged}}} & \cdots \\ \cdots & \Delta\omega_{j_2} = 100\dfrac{\omega_{j\ \text{undamaged}} - \omega_{j_2\ \text{damaged}}}{\omega_{j\ \text{undamaged}}} & \cdots \\ \cdots & \vdots & \cdots \\ \cdots & \Delta\omega_{j_p} = 100\dfrac{\omega_{j\ \text{undamaged}} - \omega_{j_p\ \text{damaged}}}{\omega_{j\ \text{undamaged}}} & \cdots \end{array} \right\} \tag{11}$$

The subscript j depends on the number of modes (d) and is selected to equal reliable experimental Eigenvalues that are excited with sufficient energy [20]. In this study, d was assigned a value of four. The matrix $z_k^{(l)}$ can be calculated based on the component of the $D_k^{(l)}$ using the updated FE model of the structure.

2.1.5. Definition of uncertainty

The uncertainty is defined by a noisy simulated MD and obtained from randomized Eigenvalues:

$$\Delta\omega_{j\ \text{noisy}} = \Delta\omega_j(1 + f\alpha), \quad j = 1, 2, \ldots, d \tag{12}$$

where f is a random and uniformly distributed number in the interval [−1, 1] and 휘 is the noise level parameter.

2.1.6. Fuzzification and the highest degree of membership for fault isolation

Here, the input variables into the FL system are defined by the Gaussian membership functions of the MDs:

$$\mu_{kj}^{(l)}(x) = e^{-0.5\left(\left(x-m_j\right)/\sigma_j\right)^2} \tag{13}$$

In this function, parameter mj is the midpoint of each column of the $\mathbf{z}_k^{(l)}$ database in the fuzzy set. The parameter σ_j is the standard deviation obtained by an algorithm that maximizes the success rate (S_R) [11]:

$$S_{R\,k_j}^{(l)} = (N_C/N_N) \times 100 \tag{14}$$

where N_C is the number of times the system classifies damage correctly and N_N is the number of noisy samples MD tested; here, N_N is considered to be equal to 1000. In Eq. (13), the x values with the highest degree of membership for slight damage, moderate damage and severe damage are obtained by considering

$$\Delta E = 25\%, \quad \Delta E = 50\% \quad \text{and} \quad \Delta E = 75\%,$$

respectively.

The fuzzy system rules are obtained for each damage class by relating the number d of the membership functions of the MDs:

$$M_k^{(l)} = \prod_{j=1}^{d} \mu_{kj}^{(l)} \tag{15}$$

The d parameter is a user defined number of measurement deltas. For the first method, the d is equal to the number of the first selected natural frequencies, which capture most of the system energy. In this study it was chosen for the first four frequencies, hence, $d = 4$ and $j = 1, 2, ..., d$ [11]. The superscript l refers to each structural member (Fig. 3) and subscript k refers to its associated damage intensity, as explain in

Section 2.1. Also, $\mu_{k\,j}^{(l)}(x)$ must be calculated using the defined database in Eq. (11). It is important to note that, for the second method, the parameters in the Eq. (15) are calculated in a different manner. The related process is explained detailed in Section 2.2.

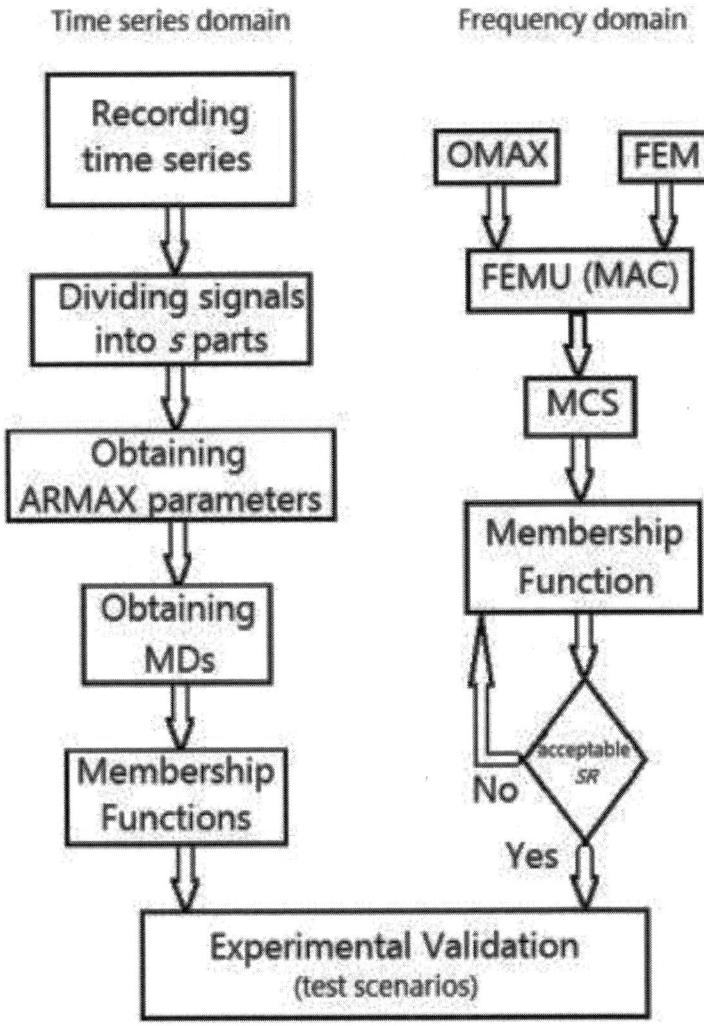

Figure. 2. Flowcharts of proposed classification methods.

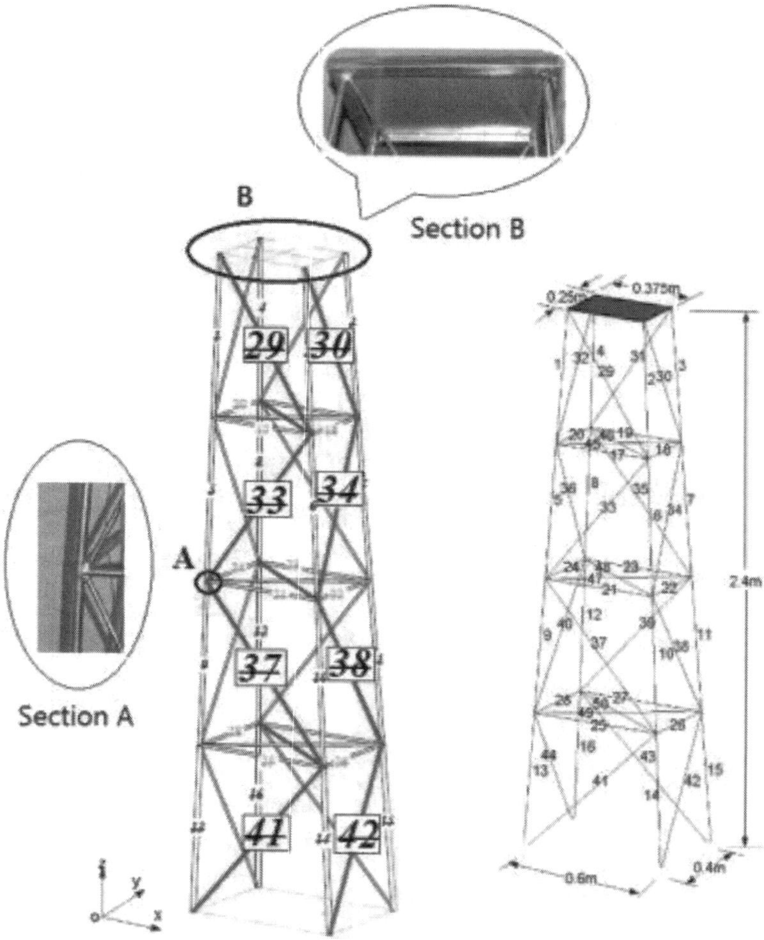

Figure. 3. The physical model description.

Finally, during the damage detection process, the class with the highest degree of membership is selected as the most likely fault.

2.2. Method based on extracted features using recorded time-series data: Black-box polynomial parametric model

This methodology is based on a strategy that can reduce the number of steps necessary for the extraction of features from the recorded data. Thus, in this case, the features have a higher heritability of the physical nature of the structure than in the previous method. Therefore, the warning changes in the physical characteristics become more traceable. Although the updating process of the first method offers some advantages [12], the conclusions about the roots of the error are more reasonable for the method based on the above-mentioned strategy. The general form of the black-box polynomial models is based on subsets of the following equation [21]:

$$A(q)y(t) = \sum_{\alpha=1}^{n_u} \frac{B_\alpha(q)}{F_\alpha(q)} u_\alpha(t - nk_\alpha) + \frac{C(q)}{H(q)} e(t) \tag{16}$$

The parameters *A*, *Bα*, *C*, *H* and *Fα* are polynomials containing the time-shift operator (q);$u\alpha$ is the αth input, *nu* is the total number of inputs and *nKα* is the αth input delay.

The model structures differ in the number of polynomials included in this structure. In the case of the ARMAX model, the autoregressive, moving average and the innovation variance parameters of the model are all expanded in proper form as follows:

$$A(q)\cdot y(t) = B(q)\cdot u(t - n_k) + C(q)\cdot e(t) \tag{17}$$

where $A(q)$, $B(q)$ and $C(q)$ are the model parameters defined below:

$$A(q) = \sum_{k=1}^{n_a} a(k)\cdot q^{-k},\ \ B(q) = \sum_{k=1}^{n_b} b(k)\cdot q^{-k},\ \ C(q) = \sum_{k=1}^{n_c} c(k)\cdot q^{-k};\ \ q^{-k}\cdot * = *\cdot(t-k) \tag{18}$$

here, *na*, *nb* and *nc* are model orders representing the number of coefficients for each polynomial, *nk* is the parameter corresponding to the input delay, and $e(t)$ is the variance of the white noise. In addition, $y(t)$ is the output signal of the dynamic system responses recorded as time-series data.

Another important step is selecting the order of the parameters and the dimension of the AR, MA and innovation variance. Various order selection criteria are available, such as the Akaike/Bayesian information criterion (AIC/BIC). The prediction error method is an extended form of the least squares method, which is used here to estimate the model parameters [21].

Note that the extracted features can be weighted by certain membership functions:

$$M^{(t)} = \left(\frac{1}{\sqrt{2\pi}}\right)^{(n_a+n_b+n_c)} \cdot \prod_{i=1}^{n_a} \frac{1}{\sigma_i^{(t)}} e^{-0.5\left(\frac{x - m_{\Delta a_i}^{(t)}}{\sigma_i^{(t)}}\right)^2} \cdot \prod_{j=1}^{n_b} \frac{1}{\sigma_j^{(t)}} e^{-0.5\left(\frac{x - m_{\Delta b_j}^{(t)}}{\sigma_j^{(t)}}\right)^2} \cdot \prod_{k=1}^{n_c} \frac{1}{\sigma_k^{(t)}} e^{-0.5\left(\frac{x - m_{\Delta}^{(t}}{\sigma_k^{(t)}}\right.} \tag{19}$$

here, σ(t) and m(t) in Eq. (19) are obtained by calculating the standard deviation and midpoint *s* of the number of measurement deltas for the *t*th test scenario. For this purpose, the recorded signals are divided into *s* equal parts, and the ARMAX parameters are obtained based on the input and output signals for the first s − 1 parts (in this study, s = 7). The measurement deltas are defined by Eq. (20):

$$\Delta a_q^{(t)}(g) = 100\frac{a_q^{(t)}(g)_{\text{undamaged}} - a_q^{(t)}(g)_{\text{damaged}}}{a_q^{(t)}(g)_{\text{undamaged}}}; \quad g = 1,2,\ldots,n_a$$

$$\Delta b_q^{(t)}(g) = 100\frac{b_q^{(t)}(g)_{\text{undamaged}} - b_q^{(t)}(g)_{\text{damaged}}}{b_q^{(t)}(g)_{\text{undamaged}}}; \quad g = 1,2,\ldots,n_b$$

$$\Delta c_q^{(t)}(g) = 100\frac{c_q^{(t)}(g)_{\text{undamaged}} - c_q^{(t)}(g)_{\text{damaged}}}{c_q^{(t)}(g)_{\text{undamaged}}}; \quad g = 1,2,\ldots,n_c \tag{20}$$

where $a_q^{(t)}$, $b_q^{(t)}$ and $c_q^{(t)}$ are obtained based on Eq. (18) and q = 1, 2, ..., s.

The flowcharts in Fig. 2 illustrate the processes of the two methods schematically.

3. DETECTION OF NONLINEARITY USING PCA AND THE CONCEPT OF SUBSPACE ANGLE

The above-mentioned damage detection methodologies were established based on linear assumptions. Therefore, it is necessary to evaluate nonlinearity errors throughout the process. For this purpose, the principal component method was adopted to use the active modes for the detection of nonlinearity in the recorded data [22]. The observation matrix is denoted R_{nm} and consists of the recorded samples, where the dimension *m* is the number of sampling points and *n* is the number of output sensors. In this study, PCA is computed from the reduced singular value decomposition (SVD) of the covariance matrix [23] based on a statistical theorem to decompose a rectangular matrix into the product of three matrices: an orthogonal matrix ***U***, in which the columns define the principal components (PCs), the diagonal matrix ***S***, and the transpose of an orthogonal matrix ***V***, as shown in Eq. (21):

$$\boldsymbol{RR}^{\mathrm{T}} = \mathbf{USV}^{\mathrm{T}} \tag{21}$$

RRT is the covariance of the observation matrix. By considering an Eigenvalue problem, the PCs can represent the vibration modes of the system. The entries in ***S*** can be restricted to the number of first singular values with a significant magnitude, *d*. Furthermore, the problem of increasing the noise levels in the PC extraction process may be considered to be another challenge for the selection of additional PCs.

The onset of nonlinearity in the dynamic system may be detected using the concept of the subspace angle, which is estimated from the reference observation set and the observation set of a current state [24]. The angles between the two subspaces are computed based on the QR decomposition concept [25]:

$$Q_O^{\mathrm{T}} Q_P = U_{OP} S_{OP} V_{OP}^{\mathrm{T}} S_{OP} = \mathrm{diag}(\cos(\theta_i)); \quad i = 1, 2, \ldots, e \tag{22}$$

where the columns of $\mathbf{Q}_O$ and $\mathbf{Q}_P$ are the orthonormal bases for the subspaces $\boldsymbol{O}_{nr}$and $\boldsymbol{P}_{ne}$, ($e<r$). There are several methods for QR decomposition, such as the Gram–Schmidt process, Householder transformations, or Givens rotations. Each has its own advantages and disadvantages.

The largest singular value is related to the largest angle characterizing the geometric difference between subspaces [26]. In this study, the above-mentioned concept is used to evaluate the effects of nonlinearity on the performance of the fault-detection methods. For comparison, the results of various tests were used to perform nonlinearity detection.

4. DESCRIPTION OF THE PHYSICAL MODEL AND TEST SETUP

For the implementation and validation of the identification methods described above, experimental modal tests were performed on a fixed jacket-type offshore platform model. The geometric dimensions of the structural members are similar to those in the model first used by Huajun et al. [6]. The general shape of the model represents a space frame with four main legs that are connected to the top deck, as shown in Fig. 3.

A physical model was constructed of stainless steel pipes that were welded together using argon arc welding to ensure proper load transfer. Metal spacers were attached separately at the middle parts of the four diagonal braces. The removal of a spacer was used to simulate the complete collapse of the brace due to damage (D = 100%). Furthermore, two replaceable diagonal bracing members were designed to be replaced with members made of different materials, i.e., aluminum, copper, and brass, with different Young's moduli to simulate stiffness reductions due to a degradation-like process. Of course, using the member replacement also changes the mass of the structure. But, it must be mentioned that the changes in the mass of these thin and slender members are negligible in comparing with the total weight. Therefore, the members with different Young's modulus are using to simulate the stiffness reductions, which can represent uniform corrosion along members. The aforementioned bracing members are illustrated in Fig. 4.

Figure. 4. Facilities for damage simulation.

The dynamic parameters of the model were acquired based on the OMAX approach. The test rig and instruments are illustrated in Fig. 5. The external excitation (based on white noise signals) was enforced by means of an electro dynamic exciter (type 4809) with a force sensor (AC20, APTech) driven by a power amplifier (model 2706), all made by Bruel&Kjaer. The instrumentation included two light uni-axial accelerometers (4508 B&K) in both the X and Y directions on each joint for response measurement and a load cell for measuring the excitation force. The frequency sampling of the test setup was chosen to be 16.385 kHz, and the frequency range was 0–200 Hz. The recorded data were sent to the PULSE [27] software package for processing.

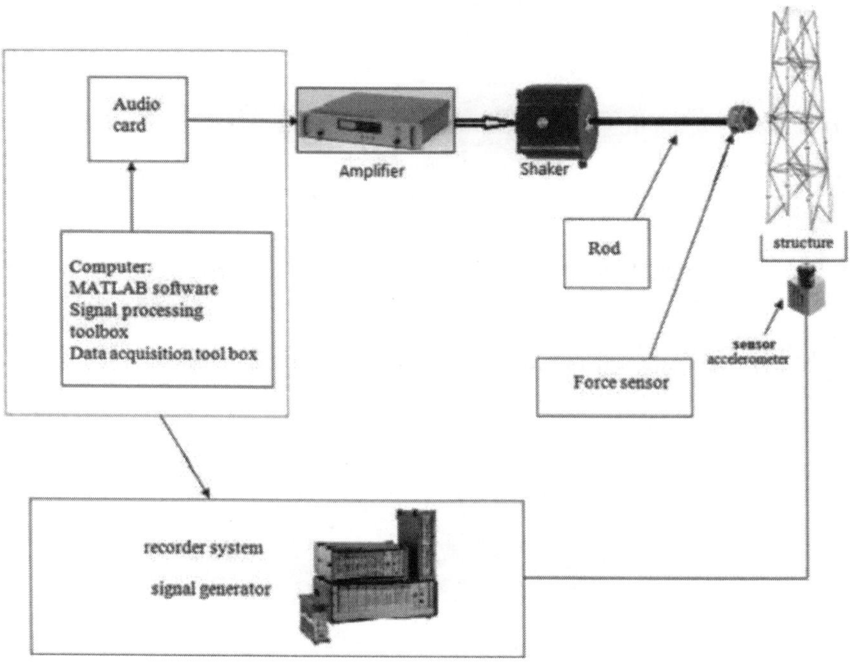

Figure. 5. The experimental test rig and instruments.

The tests were performed on the undamaged structure and then repeated in the same way for the damaged structure by reducing the stiffness of the model.

5. RESULTS AND DISCUSSION

Five experimental tests were performed based on the reported scenarios in Table 2. The application of the mounted accelerometer sensors for damage detection purposes has been evaluated when the number of available sensors is small. Two different methods have been proposed; the first method uses the modal parameter in the frequency domain, and the second one uses the extracted features from output time-series signals. The advantages and disadvantages of these methods are compared with each other, and the effects of the activation of nonlinear dynamic behavior and noise parameter on the efficiency of the mare investigated. The observed results are presented in this section and discussed as follows.·

Table 1. Initial values of the elasticity modulus for the complete FE model after updating.

Member	29	30	33	34	37	38	41	42
E (GPa)	207	207	194	181	194	194	181	207

Table 2. Experimental damage scenarios and the efficiencies of the two methods.

Case number	Substituted member	Replacement material	D (%)	Damage intensity	Method efficiency	
					Freq. domain (FLS)	Time domain (ARMAX)
1	42	Copper	41	Moderate	Successful	Successful
2	42	Aluminum	65	Severe	Successful	Successful
3	38	Aluminum	65	Severe	Successful	Successful
4	32	Eliminating	100	Catastrophic	*Unsuccessful*	*Unsuccessful*
5	41	Eliminating	100	Catastrophic	Successful	Successful

5.1. FE model updating and experimental modal features

The ME'scope software was used to obtain the experimental modal parameters by polynomial curve fitting of the frequency response functions (FRFs). The data required for calculating the FRFs were recorded by sensors that were fixed on the physical model joints. Because there were more desired points for measurement (i.e., the 20 joints of the model) than the number of available channels and accelerometers, the measurements were performed in 20 steps. The database of acquired parameters consists of the first four experimental natural frequencies of the physical model.

The ANSYS FE package was employed to obtain the numerical modal parameters described in Section 2. The Young's modulus, Poisson ratio and density were 207 GPa, 0.3 and 7850 kg/m^3, respectively. The resulting numerical parameters were somewhat inconsistent with the experimental values. For a damage detection process, the experimental modal results are far more acceptable, and the numerical model was updated according to the experimental results to minimize these differences, as explained in Section 2.1. The first four natural frequencies are shown in Fig. 6, before and after the updating process (these lines were drawn only to make the visual inspection more clear for pre- and post-updated points.). Their sensitivity to Young's modulus is also illustrated in this figure when E is changed by 75% (severe damage) simultaneously for all vertical bracing members in one span of the model.

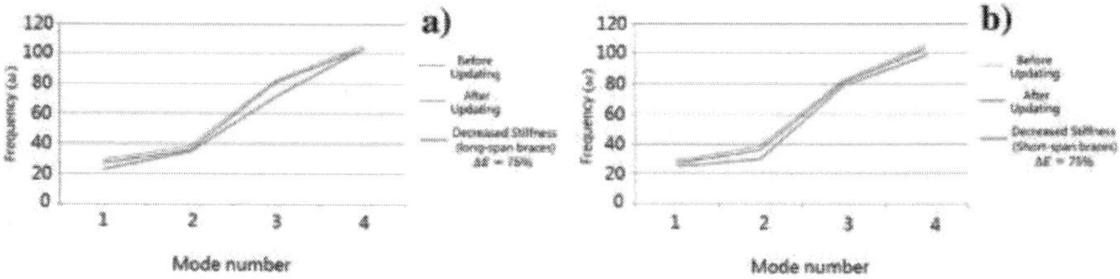

Figure. 6. Sensitivity of the first four natural frequencies to variation of the modulus of elasticity on vertical bracing members: (a) long span, (b) short span.

The sensitivity between the stiffness of the vertical bracing members 29, 30, 33 and 42 and their natural frequencies are reported in Fig. 7.

Using a sensitivity analysis, the stiffness of the bracing members were identified as adjustable parameters for the updating process, and the final values for the complete numerical model were determined, as listed in Table 1.

The updated FE model was used to train the algorithm for the first damage recognition method based on the FL system. The MDs (fuzzy variables), mean, standard deviations and membership functions were obtained from a Monte Carlo simulation to avoid the uncertainty associated with the variables. The detailed results are described elsewhere[11].

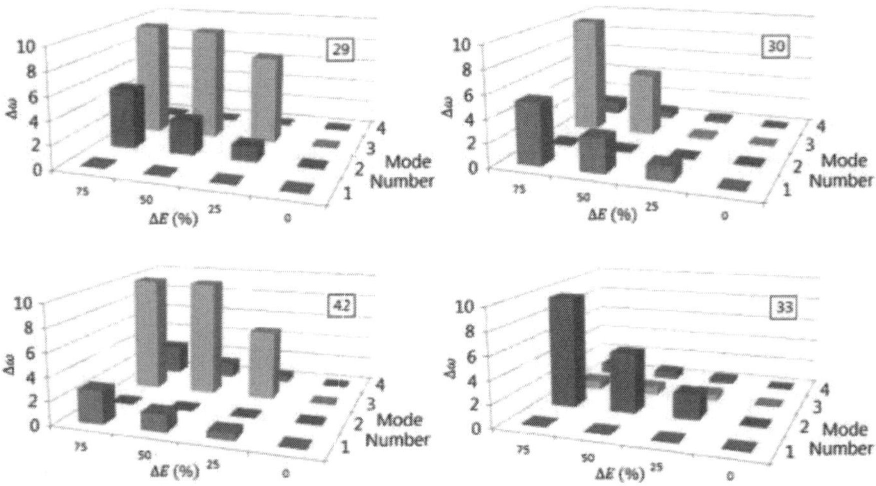

Figure . 7. Sensitivity of the first four frequencies and the damage percentages in bracing members 29, 30, 33 and 42.

5.2. Black-box polynomial parametric model

In this study, experimental tests were performed based on certain damage scenarios. The number and setup of the sensors must be considered key factors in the success of these methods, which also depends on the number of the PCs that are used for the calculation of the subspace angles. We should note that the detection process may deteriorate with the use of many PCs due to the noise that perturbs the procedure. Based on the observed results in the model updating process, the use of four PCs returned good cumulative energies, allowing for a suitable representation of the observation matrix. Therefore, four PCs were considered in the subspaces, and consequently, four accelerometers were installed on two different joints. The accelerometers were fixed in two direction sat each joint, as shown in Fig. 8.

Random forces were applied to capture the random responses of the model. The samples of the compressed acceleration data along with their corresponding power spectral densities (PSDs) for the normal case and Damage Scenario 3 are shown in Figs. 9 and 10, respectively. Various practical considerations must be made to collect high-quality data in an experimental test. For this purpose, the sampled signals must contain periodic repetitions of the measured data to yield a proper representation in the frequency domain [16]. The test setup must be selected so that all modes of interest are considered. Additionally, the coherence function must be considered to be another important parameter, and an attempt should be made to obtain a relatively flat and well-behaved function.

Figure. 8. Installed accelerometers and experimental measurements on the model joints.

The second identification methodology was developed based on the features that are extracted directly from the output time-series signals, unlike the previous method. Therefore, the steps necessary for the extraction of these features from the original recorded data are reduced, hence it can authenticate the originality of data. Therefore, the changes in the physical parameters become more traceable. Furthermore, the first method, which is based on a model updating process, requires more computational time.

The second approach is based on the stochastic ARMAX parametric models. Many advantages are found in parametric time-domain methods [28], such as specifying the physics of the problem with a limited number of parameters, improvement in tracking the time-varying dynamics, flexibility in both simulation and prediction goals, and flexibility in fault diagnosis.

In this study, the recorded data are divided into seven equal parts during the required calculation process, as shown in Fig. 11. The means, standard deviations and membership functions were obtained for the ARMAX parameters using the first six parts based on the process explained in Section 2.2. The efficiency of the model for each test scenario was validated using the seventh part.

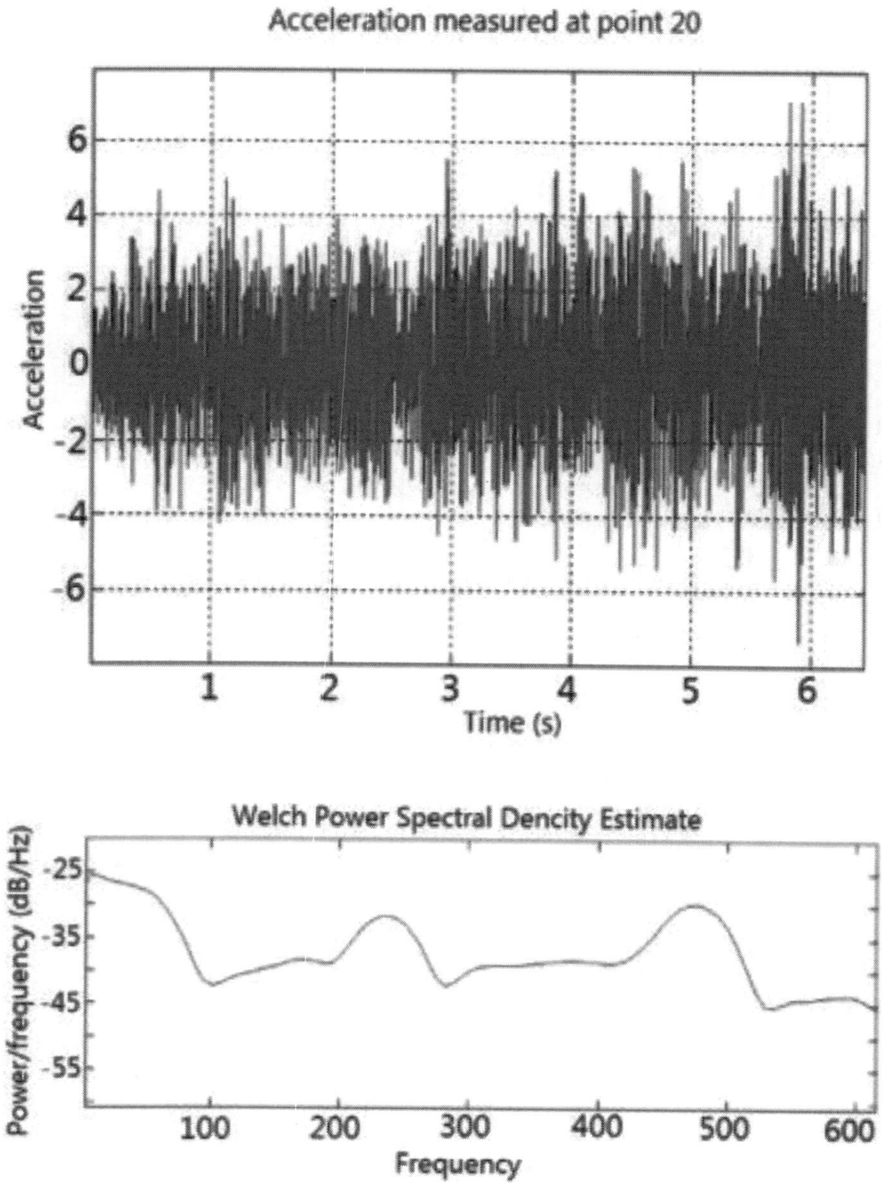

Figure. 9. The measured compressed acceleration data and the corresponding PSD for the intact pattern.

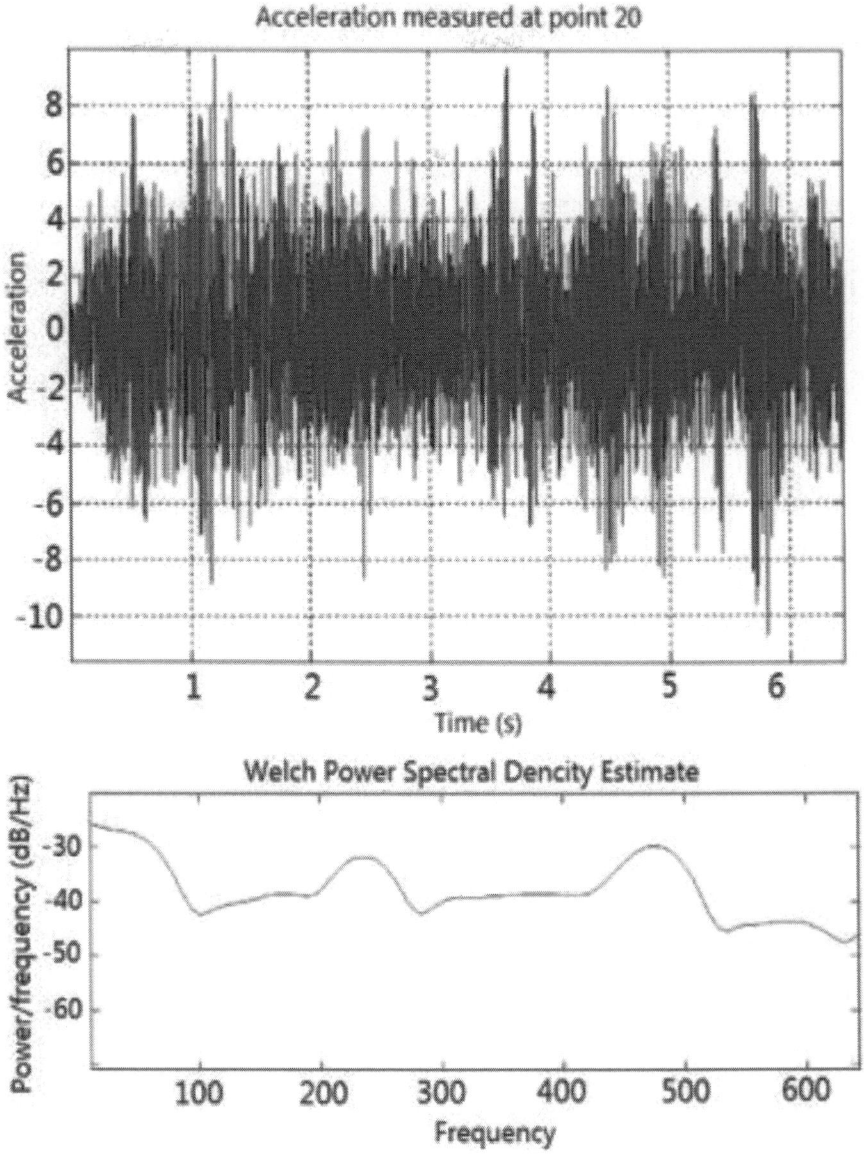

Figure. 10. The measured compressed acceleration data and the corresponding PSD for Damage Scenario 3.

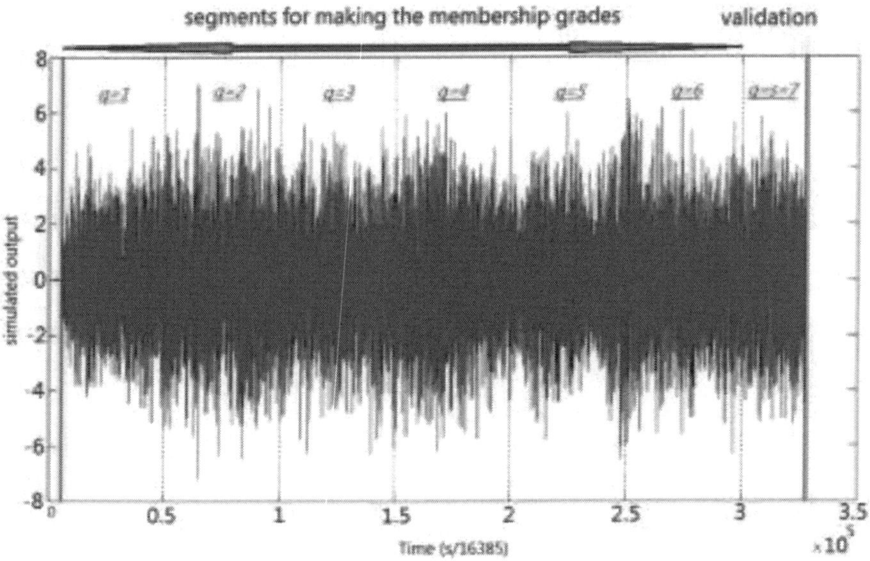

Figure. 11. Division of the signals for calculation of the membership grade and validation of the ARMAX model.

5.3. Classification results

Five damage scenarios were tested in the form of stiffness reductions by changing certain members of the model or removing them completely. The details of these scenarios, including member numbers, substituted materials and percentage damage parameters, are listed in Table 2.

The performance of the FL system method was evaluated using the training and classification data, as described in previous sections. These results imply the efficiency of the technique for diagnosing the considered empirical damage. Nevertheless, the detection failed in case number 4, as shown in Table 2.

To investigate the efficiency of the second method, the signals were divided into seven parts. The order of the ARMAX model was selected based on the Akaike's Bayesian information criterion (AIC/BIC). Thus, the order was selected based on a criterion that is a measure of the goodness of fit for estimated statistics. The orders of the AR, MA, and X parts and the variance of the model were determined to be $n_a = 2$, $n_b = 4$ and $n_c = 4$, respectively. Finally, the membership functions were calculated using the ARMAX parameters related to the first six parts of the signal using Eq. (19). The seventh part was used to validate the model.

The measured and simulated time series from the modeling of the validation signal for case number 3 are compared in Fig. 12. The validation results demonstrate that he parametric model is accurate and suitable for this test scenario.

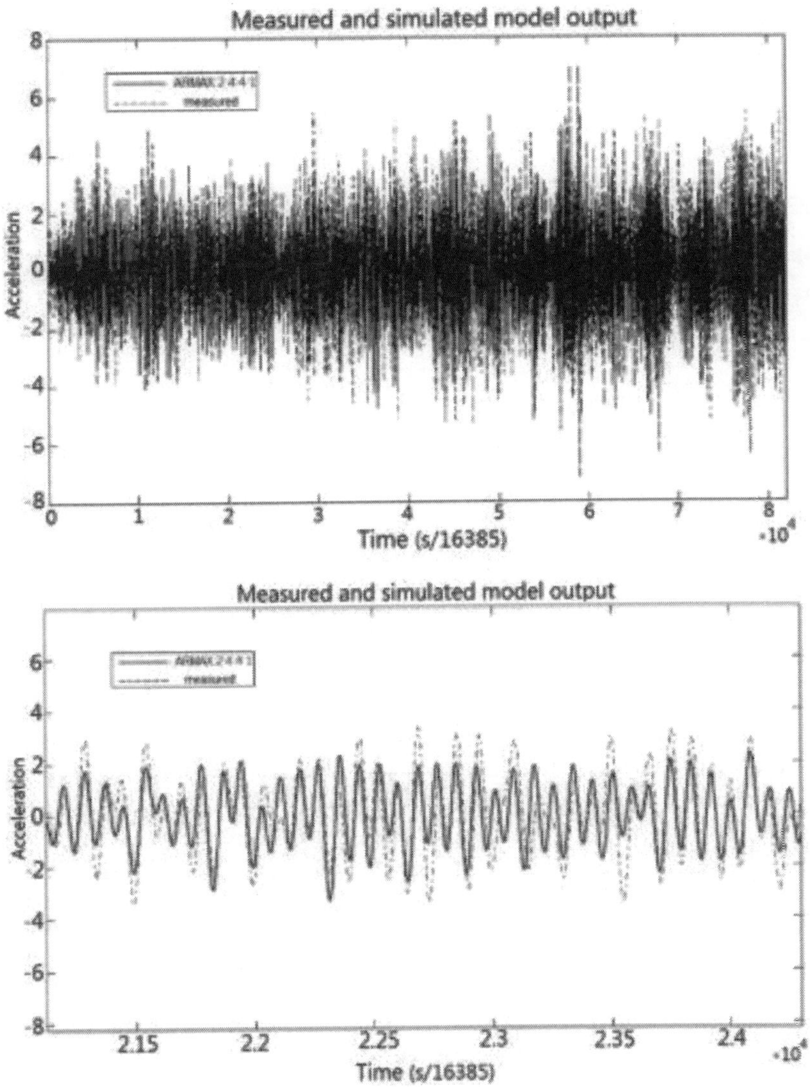

Figure. 12. Measured output signals and simulation by the ARMAX model for damage scenario 3.

The performance of the second method was evaluated for the five damage scenarios, as for the first method. The value of each membership grade was calculated by inputting the related ARMAX parameters into the variable *x* of Eq. (19). In a successful process, the parameters must maximize their associated membership grades. The results in Table 2show that this method failed in the diagnosis of the fourth case, as with the first method.

Fig. 13 illustrates the measured and simulated time series for case number 4. As shown here, the method is not accurate for case number 4 compared with the others.

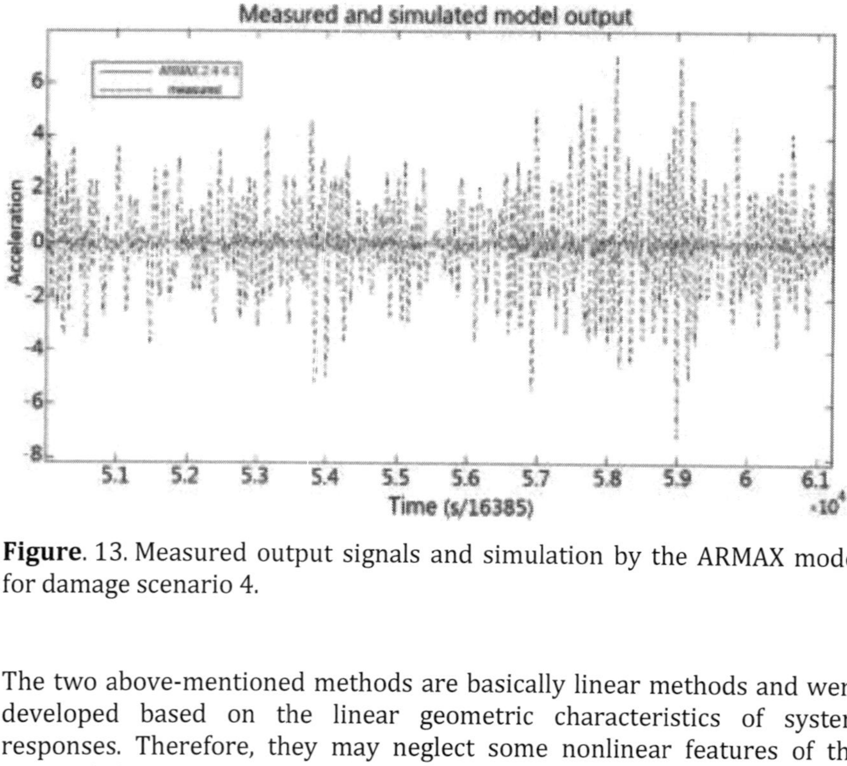

Figure. 13. Measured output signals and simulation by the ARMAX model for damage scenario 4.

The two above-mentioned methods are basically linear methods and were developed based on the linear geometric characteristics of system responses. Therefore, they may neglect some nonlinear features of the system behavior. To investigate the failed test scenario, the nonlinearity in the dynamic system was evaluated. The PCA and concept of subspace angle were adopted to detect nonlinearity in the recorded data. One advantage of this approach is that it can be used directly as a criterion for damage classification by monitoring the angular coherence between the subspaces of the intact condition and an observed damage scenario.

As the observation matrix was constructed based on the four response vectors, four PCs were used to calculate the PCA and detect the onset of nonlinearity based on the subspace angle. Time samples of 0.6 s were analyzed, and frequency sampling of the measured data was reduced to 160 Hz to decrease the dimension of the subspace matrices and hence reduce the computational time. The results are reported in Fig. 14 for all five scenarios. Good agreement was observed with the results in Table 2. The detection index gave a large angle value for case 4. In this case, the nonlinear state can be well distinguished in the recorded system responses.

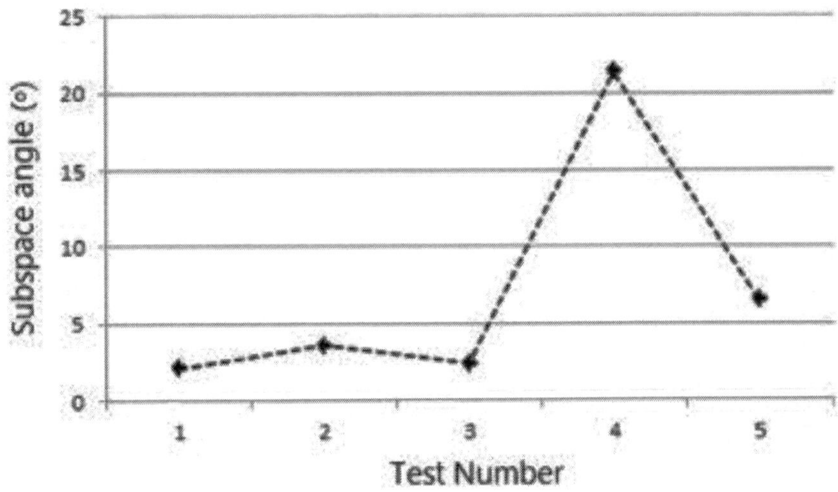

Figure. 14. PCA nonlinearity detection based on the subspace angle between the intact and damaged test scenarios.

The effect of noise in the extracted data was examined by adding 5% noise to each measured time series. The noisy signals were simulated with a randomly (Gaussian) distributed number with a mean of 0 and a standard deviation of 1·[29] (where the noise level parameter $\alpha = 0.05$). The results for 10 noisy data sets are shown in Fig. 15. Here, the nonlinearities are well distinguished, as in Fig. 14. The observed results indicate that the inefficiency of the proposed methods is related to the level of nonlinearity in the dynamic behavior of the measured signals.

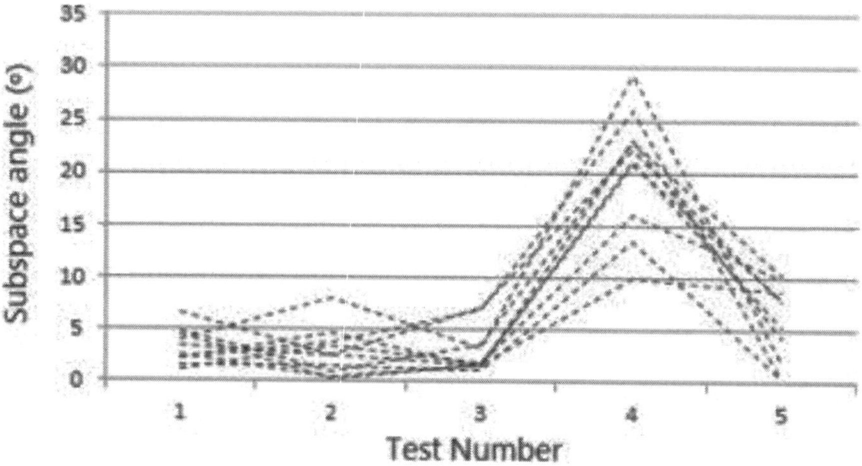

Figure15. PCA nonlinearity detection for damage test scenarios with 5% noise added to the measured signals.

6. CONCLUSION

Two classification algorithms for classifying damages in offshore jacket-type structures are presented and were inspired separately by experimental modal analysis and time-capture data processing. The offshore structures are installed in the ocean and subjected to hydrodynamic loads. As the considered feature sets and methodologies of this study are independent of the excitation types, the proposed techniques are adaptable for different input forces. Therefore, the inherent structural dynamic output response must be considered to be a key point of these methods. The responses can be recorded for any type of external excitation, such as hydrodynamic loads, machinery vibrations within the structures, seismic loads, or wave action. The proposed methods were validated using the vibration characteristics in dry conditions.

The efficiency of the methods was investigated and compared. The validation results show that these methods are suitable for damage classification in the considered structure. The first method, which is a combined algorithm consisting of an FL system and a model updating process, offers the advantage of allowing for damage location and intensity calculation. However, it generally requires more computational time than the second method.

The advantages of using the second parametric time-domain method are that the physics of the problem can be specified by a limited number of parameters, tracking of the time-varying dynamics is improved, and it allows for flexibility in both simulation and prediction goals.

However, with both methods, damage detection failed completely in one of the cases. The principal component method and the concept of subspace angle were used to evaluate the nonlinearity in this dynamic system. The observed results imply that the efficiency of the proposed methods is related to the onset of geometric nonlinear behavior in the measured signals. The problem can be addressed by adjusting sensor positioning and improving the quality of the acquired signals. In addition, enhancement of the described techniques based on non-stationary and nonlinear methods, such as the TARMAX model [30] or KPCA analysis [31], can be considered as other solutions in future research.

REFERENCES

1. Doebling SW, Farrar CR, Prime MB. A summary review of vibration based damage identification methods. J Shock Vib Dig 1998;30:91–105.
2. Mangal L, Idichandy VG, Ganapathy C. Structural monitoring of offshore platforms using impulse and relaxation response. J Ocean Eng 2001;28:689–705.
3. Nichols JM. Structural health monitoring of offshore structures using ambient excitation. J Appl Ocean Res 2003;25:101–14.
4. Elshafey A, Haddara Mahmoud R, Marzouk H. Damage detection in offshore structures using neural networks. J Mar Struct 2010;23:131–45.
5. Elshafey A, Haddara MR, Marzouk H. On the use of mode shapes in the identification of structural damage. Proceedings of the ASME 2010 29th international conference on ocean, offshore and arctic engineering OMAE2010, Shanghai, China; June 6–11, 2010.
6. Li Huajun, Wang Shuqing, Yang Hezhen. Modal strain energy decomposition method for damage detection of an offshore structure using modal testing information. In: Third Chinese German joint symposium on coastal and ocean engineering. Tainan: National Cheng Kung University; November 8–16, 2006.
7. Junfei Liu, Huajun Li, Yong Wang, Xingtao Hu, Modal parameters identification of offshore platform structures using HHT method, Proceedings of the nineteenth international offshore and polar engineering conference, Osaka, Japan; June 21–26, 2009.

8. Ney Roitman, Anderson S.M. Gadéa, Carlos Magluta, Damage identification in offshore platforms using frequency response functions, Proceedings of OMAE04 23rd international conference on offshore mechanics and arctic engineering, Vancouver, British Columbia, Canada; June 20–25, 2004.

9. Farrar Charles R, Worden Keith. An introduction to structural health monitoring. Phil Trans R Soc A 2007;365:303–15.

10. Reynders E, Teughels A, DeRoeck G. Finite element model updating and structural damage identification using OMAX data. J Mech Syst Signal Process 2010;24:1306–23.

11. Mojtahedi A, LotfollahiYaghin MA, Hassanzadeh Y, Ettefagh MM, Aminfar MH, Aghdam AB. Developing a robust SHM method for offshore jacket platform using model updating and fuzzy logic system. J Appl Ocean Res 2011;33:398–411.

12. Modak SV, Kundra TK, Nakra BC. Comparative study of model updating methods using experimental data. J Comp Struct 2000;80:437–47.

13. Papatheou Evangelos, Manson Graeme, Barthorpe Robert J, Worden Keith. The use of pseudo-faults for novelty detection in SHM. J Sound Vib 2010;329:2349–66.

14. Bayraktar Alemdar, Sevim Bar, Altun Ahmet Can, Türker Temel. Effect of the model updating on the earthquake behavior of steel storage tanks. J Const Steel Res 2010;66:462–9.

15. Peeters M, Kerschen G, Golinval JC. Modal testing of nonlinear vibrating structures based on nonlinear normal modes: experimental demonstration. J Mech Syst Signal Process 2011;25:1227–47.

16. Avitable P. Experimental modal analysis; a simple non-mathematical presentation. J Sound Vib 2001. Draft document.

Index